D0350224

A Guide to the End of the World

A Guide to the End of the World

EVERYTHING YOU NEVER WANTED TO KNOW

Bill McGuire

OXFORD
UNIVERSITY PRESS

Great Clarendon Street, Oxford OX2 6DP

Oxford University Press is a department of the University of Oxford.
It furthers the University's objective of excellence in research, scholarship,
and education by publishing worldwide in

Oxford New York

Auckland Bangkok Buenos Aires Cape Town Chennai
Dar es Salaam Delhi Hong Kong Istanbul Karachi Kolkata
Kuala Lumpur Madrid Melbourne Mexico City Mumbai Nairobi
São Paulo Shanghai Singapore Taipei Tokyo Toronto

with an associated company in Berlin

Oxford is a registered trade mark of Oxford University Press
in the UK and in certain other countries

Published in the United States
by Oxford University Press Inc., New York

First published 2002

British Library Cataloguing in Publication Data
Data available

Library of Congress Cataloging in Publication Data

Data available

ISBN 0–19–280297–6

1 3 5 7 9 10 8 6 4 2

Typeset in New Baskerville
by RefineCatch Limited, Bungay, Suffolk
Printed in Spain by
Book Print S.L., Barcelona

For Jetsam, Driftwood, and the late, lamented Flotsam

Foreword—where will it all end?

Que será, será
Whatever will be will be
The future's not ours to see
Que será, será

[(Jay Livingston and Ray Evans)]

The big problem with predicting the end of the world is that, if proved right, there can be no basking in glory. This has not, however, dissuaded armies of Cassandras from predicting the demise of our planet or the human race, only to expire themselves without the opportunity to proclaim 'I told you so'. To somewhat adapt the words of the great Mark Twain, the death of our race has been greatly exaggerated. The big question is, however, how long will this continue to be the case?

In answer, it would be perfectly reasonable to say that of course the world is going to end—in about five billion years time when our Sun finally runs out of fuel and swells to become a bloated red giant that burns the Earth to a cinder. A fervent eschatologist, however, would undoubtedly contest this, launching into an enthusiastic account of the many alternative and imaginative ways in which our world and our race might meet its end, of which disease, warfare, natural catastrophe, and exotic physics experiments gone wrong are but a selection. Given the current state of the planet you might be forgiven for having second thoughts following such a litany—perhaps, after all, we will face 'doom soon' as John Leslie succinctly put it in *his* book *The End of the World*, rather than 'doom deferred'. Against a background of accelerating global warming, exploding population, and reborn superpower militarism, it may indeed be more logical for us to speculate that the human race's great adventure is about to end, rather than persist far into the future and across the vastness of galactic space.

Somewhat worryingly, Cambridge cosmologist Brandon Carter has developed an argument that supports, probabilistically, this very thesis. His 'doomsday argument' goes like this. Assuming that our race grows and persists for millions or even billions of years, then those of us alive today must belong to the infinitesimally small fraction of humans living in the earliest light of our race's dawn. This, Carter postulates, is statistically unlikely in the extreme. It is much more probable that we are alive at the same time as, say, 10 per cent of the human race. This is another way of saying that humans will cease to exist long before they have any chance to spread across space in any numbers worth talking about.

John Leslie illustrates this argument along these lines. Imagine your name is in a lottery draw, but you don't know how many other names there are. You have reason to believe, however, that there is a 50 per cent chance that the total number is a thousand and an equal probability that the total is ten. When the tickets are drawn, yours is one of the first three. Now, there can be few people who, in such circumstances, would believe that the draw contained a thousand rather than ten tickets.

If the doomsday argument is valid—and it has withstood some pretty fierce attacks from a number of intellectual heavyweights—then we may have only a few centuries' respite before one Nemesis or another obliterates our race, our planet, or both. Despite nearly a quarter of a century in the 'doom and disaster' business, however, I can't help being at least a little optimistic. Wiping out 6 billion or more people at a stroke will not be easy, and many of the so-called 'end of the world' scenarios are actually no such thing, but would simply result—at worst—in a severe fall in human numbers and/or the reduction of our global, technological civilization to something far simpler and more parochial—at least for a time. Personally, therefore, I am open-minded about what Stephen Baxter calls in his recent novel *Manifold Time* the 'Carter Catastrophe'. There is no question that the human race or its descendants must eventually succumb to oblivion, but that time may yet be a very long way off indeed.

This might be a good point to look more carefully at just what we understand by 'the end of the world', and how I will be treating the concept in this book. To my thinking, it may be interpreted in four different ways: (i) the wholesale destruction of the planet and the race, which will certainly occur if all the human eggs remain confined to our single terrestrial basket when our Sun 'goes nova' five billion years hence; (ii) the loss of our planet to some catastrophe or another, but the survival of at least some elements of our race on other worlds; (iii) the obliteration of the human race but the survival of the planet, due perhaps to some virulent and inescapable disease; and (iv) the end of the world *as we know it.* It is on this final scenario that I will be focusing here, and the main thrust of this book will address global geophysical events that have the potential to deal our race and our global technological society a severe, if not lethal blow. Natural catastrophes on a scale mighty enough to bring to an end our familiar world. I will not concern myself with technological threats such as those raised by advances in artificial intelligence and robotics, genetic engineering, nano-technology, and increasingly energetic high-energy physics experiments. Neither will I address—barring global warming—attempts by some of the human race to reduce its numbers through nuclear, biological, or chemical warfare. Instead I want to introduce you to some of the very worst that nature can throw at us, either solely on its own account or with our help.

Although often benign, nature can be a terrible foe and mankind has fought a near-constant battle against the results of its capriciousness—severe floods and storms, devastating earthquakes, and cataclysmic volcanic eruptions. So far, however, we have been quite fortunate, and our civilization has grown and developed against a backdrop of relative climatic and geological calm. The omens for the next century and beyond are, however, far from encouraging. Dramatic rises in temperature and sea level in coming decades induced by greenhouse gases—in combination with ever-growing populations—will without doubt result in a huge increase in the number and intensity of natural disasters. Counter-intuitively, some parts of the planet may even end up getting much colder and the UK, for example, could—in a century or two—be freezing in Arctic conditions as the Gulf Stream weakens. And what exactly happened to the predicted new Ice Age anyway? Has the threat gone away with the onset of anthropogenic (man-made) global warming or are the glaciers simply biding their time?

Although rapid in geological terms, climate change is a slow-onset event in comparison with the average human lifespan, and to some extent at least its progress can be measured and forecast. Much more unexpected and difficult to predict are those geological events large enough to devastate our entire society and which we have yet to experience in modern times. These can broadly be divided into extraterrestrial and terrestrial phenomena. The former involve the widely publicized threat to the planet arising from collisions with comets or asteroids. Even a relatively small, one-kilometre object striking the planet could be expected to wipe out around a quarter of the Earth's population.

The potential for the Earth itself to do us serious harm is less widely documented, but the threat of a major natural catastrophe arising from the bubbling and creaking crust beneath our feet is a real and serious one. Three epic events await us that have occurred many times before in our planet's prehistory, but which we have yet to experience in historic time. The last volcanic *super-eruption* plunged the planet into a bitter *volcanic winter* some 73,500 years ago, while little more than 100,000 years ago gigantic waves caused by a collapsing Hawaiian volcano mercilessly pounded the entire coastline of the Pacific Ocean. Barely a thousand years before the birth of Christ, and again during the Dark Ages, much of eastern Europe and the Middle East was battered by an earthquake *storm* that levelled once great cities over an enormous area. There is no question that such *tectonic* catastrophes will strike again in our future, but just what will be their effect on our global, technology-based society? How well we will cope is difficult to predict, but there can be little doubt that for most of the inhabitants of Earth, things will take a turn for the worse.

Living on the most active body in the solar system, we must always keep in our minds that we exist and thrive only by geological accident. As I will address in Chapter 4, recent studies on human DNA have revealed that our race came within a hair's breadth of extinction following the last super-eruption 73,500 years ago, and if we had been around 65 million years ago when a ten-kilometre asteroid struck the planet we would have vanished with the dinosaurs. We must face the fact that, as long as we are all confined to a single planet in a single solar system, the long-term survival of our race is always going to be tenuous. However powerful our technologies become, as long as we remain in Earth's cradle we will always be dangerously exposed to nature's every violent whim. Even if we reject the 'doom soon' scenario, it is likely that our progress as a race will be continually impeded or knocked back by a succession of global natural catastrophes that will crop up at irregular intervals as long as the Earth exists and we upon it. While some of these events may bring to an end the world as we know it, barring another major asteroid or comet impact on the scale of the one which killed the dinosaurs, the race is likely to survive and, generally, to advance. At some point in the future, therefore, we will begin to move out into space—first to our sibling worlds and then to the stars. In the current inward-looking political climate it is impossible to say when a serious move into space will happen, but happen it will and when it does the race will breathe a collective sigh of relief. At last some of our eggs will be in a different basket. What happens next is anyone's guess. As this book will show, when it comes to geophysics, what will be, will be.

Bill McGuire

Hampton, England
August 2001

Contents

List of Illustrations

A Very Short Introduction to the Earth

Danger: nature at work

We are so used to seeing on our television screens the battered remains of cities pounded by earthquakes or the thousands of terrified refugees escaping from yet another volcanic blast that they no longer hold any surprise or fear for us, insulated as we are by distance and a lack of true empathy. Although not entirely immune to disaster themselves, the great majority of citizens fortunate enough to live in prosperous Europe, North America, or Oceania view great natural catastrophes as ephemeral events that occur in strange lands far, far away. Mildly interesting but only rarely impinging upon a daily existence within which a murder in a popular soap opera or a win by the local football team holds far more interest than 50,000 dead in a Venezuelan mudslide. Remarkably, such an attitude even prevails in regions of developed countries that are also susceptible to volcanic eruptions and earthquakes. Talk to the citizens of Mammoth in California about the threat of their local volcano exploding into life, or to the inhabitants of Memphis, Tennessee, about prospects for their city being levelled by a major quake, and they are likely to shrug and point out that they have far more immediate things to worry about. The only explanation is that these people are in denial. They are quite aware that terrible disaster *will* strike at some point in the future—they just can't accept that it might happen to them or their descendants.

When it comes to natural catastrophes on a global scale such an attitude is virtually omnipresent, pervading national governments, international agencies, multinational trading blocks, and much of the scientific community. There is some cause for optimism, however, and in one area, at least, this has begun to change. The threat to the Earth from asteroid and comet impacts is now common knowledge and the race is on to identify all those Earth-approaching asteroids that have the potential to stop the development of our race in its tracks. Thanks to recent widely publicized television documentaries shown in the UK and United States, the added threats of volcanic super-eruptions and giant tsunamis have now also begun to reach an audience wider than the tight groups of scientists that work on these rather esoteric phenomena.

In fact, the Earth is an extraordinarily fragile place that is fraught with danger: a tiny rock hurtling through space, wracked by violent movements of its crust and subject to dramatic climatic changes as its geophysical and orbital circumstances vary. Barely 10,000 years after the end of the Ice Age, the planet is sweltering in some of the highest temperatures in recent Earth history. At the same time, overpopulation and exploitation are dramatically increasing the vulnerability of modern society to natural catastrophes such as earthquakes, floods, and volcanic eruptions. In this introductory chapter, current threats to the planet and its people are examined as a prelude to consideration of the bigger threats to come.

The Earth is the most dynamic planet in our solar system, and it is this dynamism that has given us our protective magnetic field, our atmosphere, our oceans, and ultimately our lives. The very same geophysical features that make the Earth so life-giving and preserving also, however, make it dangerous. For example, the spectacular volcanoes that in the early history of our planet helped to generate the atmosphere and the oceans have in the last three centuries wiped out a quarter of a million people and injured countless others. At the same time, the rains that feed our rivers and provide us with the potable water that we need to survive have devastated huge tracts of the planet with floods that in recent years have been truly biblical in scale. In any single year since 1990 perhaps 20,000 were killed and tens of millions affected by raging floodwaters, and in 1998 major river floods in China and Bangladesh led to misery for literally hundreds of millions of their inhabitants. I could go on in the same vein, describing how lives made enjoyable by a fresh fall of snow are swiftly ended when it avalanches, or how a fresh breeze that sets sailing dinghies skimming across the wave tops can soon transform itself into a wailing banshee of terrible destruction—but I think you get the picture. Nature provides us with all our needs but we must be very wary of its rapidly changing moods.

The Earth: a potted biography

The major global geophysical catastrophes that await us down the line are in fact just run-of-the-mill natural phenomena writ large. In order to understand them, therefore, it is essential to know a little about the Earth and how it functions. Here, I will sashay through the 4.6 billion years of Earth history, elucidating along the way those features that make our world so hazardous and our future upon it so precarious. To begin, it is sometimes worth pondering upon just how incredibly old the Earth is, if only to appreciate the notion that just because we have not experienced a particular natural catastrophe before does not mean it has never happened, nor that it will not happen again. The Earth has been around just about long enough to ensure that anything nature can conjure up it already has. To give a true impression of the great age of our planet compared to that of our race, perhaps I can fall back on an analogy I have used before. Imagine the entirety of Earth's history represented by a team of runners tackling the three and a half laps of the 1,500 metres. For the first lap our planet would be a barren wasteland of impacting asteroids and exploding volcanoes. During the next the planet would begin to cool, allowing the oceans to develop and the simplest life forms to appear. The geological period known as the *Cambrian*, which marked the real explosion of diverse life forms, would not begin until well after the bell has rung and the athletes are hurtling down the final straight of the last lap. As they battle for the tape, dinosaurs appear and then disappear while the leaders are only 25 metres from the finish. Where are we? Well, our most distant ancestors only make an appearance in the last split-second of the race, just as the exhausted winner breasts the tape.

Since the first single-celled organisms made their appearance billions of years ago, within sweltering chemical soups brooded over by a noxious atmosphere, life has struggled precariously to survive and evolve against a background of potentially lethal geophysical phenomena. Little has changed today, except perhaps the frequency of global catastrophes, and many on the planet still face a daily threat to life, limb, and livelihood from volcano, quake, flood, and storm. The natural perils that have battered our race in the past, and which constitute a growing future threat, have roots that extend back over 4 billion years to the creation of the solar system and the formation of the Earth from a disc of debris orbiting a primordial Sun. Like our sister planets, the Earth can be viewed as a lottery jackpot winner; one of only nine chunks of space debris out of original trillions that managed to grow and endure while the rest annihilated one another in spectacular collisions or were swept up by the larger lucky few with their stronger and more influential gravity fields. This sweeping-up process—known as *accretion*—involved the Earth and other planets adding to their masses through collisions with other smaller chunks of rock, an extremely violent process that was mostly completed—fortunately for us—almost 4 billion years ago. After this time, the solar system was a much less cluttered place, with considerably less debris hurtling about and impacts on the planets less ubiquitous events. Nevertheless, major collisions between the Earth and *asteroids* and *comets*—respectively rocky and icy bodies that survived the enthusiastic spring cleaning during the early history of the solar system—are recognized throughout our planet's geological record. As I will discuss in Chapter 5, such collisions have been held responsible for a number of mass extinctions over the past half a billion years, including that which saw off the dinosaurs. Furthermore, the threat of asteroid and comet impacts is still very much with us, and over 300 *Potentially Hazardous Asteroids* (or PHAs) have already been identified that may come too close for comfort.

The primordial Earth would have borne considerably more resemblance to our worst vision of hell than today's stunning blue planet. The enormous heat generated by collisions, together with that produced by high concentrations of radioactive elements within the Earth, would have ensured that the entire surface was covered with a churning magma ocean, perhaps 400 kilometres deep. Temperatures at this time would have been comparable with some of the cooler stars, perhaps approaching 5,000 degrees Celsius. Inevitably, where molten rock met the bitter cold of space, heat was lost rapidly, allowing the outermost levels of the magma ocean to solidify to a thin crust. Although the continuously churning currents in the molten region immediately below repeatedly caused this to break into fragments and slide once again into the maelstrom, by about 2.7 billion years ago more stable and long-lived crust managed to develop and to thicken gradually. *Convection currents* continued to stir in the hot and partially molten rock below, carrying out the essential business of transferring the heat from radioactive sources in the planet's deep interior into the growing rigid outer shell from where it was radiated into space. The disruptive action of these currents ensured that the Earth's rigid outer layer was never a single, unbroken carapace, but instead comprised separate rocky *plates* that moved relative to one another on the backs of the sluggish convection currents.

As a crust was forming, major changes were also occurring deep within the Earth's interior. Here, heavier elements—mainly iron and nickel—were slowly sinking under gravity towards the centre to form the planet's metallic core. At its heart, a ball made up largely of solid iron and nickel formed, but pressure and temperature conditions in the outer core were such that this remained molten. Being a liquid, this also rotated in sympathy with the Earth's rotation, in the process generating a magnetic field that protects life on the surface by blocking damaging radiation from space and provides us with a reliable means of navigation without which our pioneering ancestors would have found exploration—and returning home again—a much trickier business.

For the last couple of billion years or so, things have quietened down considerably on the planet, and its structure and the geophysical processes that operate both within and at the surface have not changed a great deal. Internally, the Earth has a threefold structure. A crust made up of low-density, mainly silicate, minerals incorporated into rocks formed by volcanic action, sedimentation, and burial; a partly molten *mantle* consisting of higher-density minerals, also silicates, and a composite core of iron and nickel with some impurities. Ultimately, the hazards that constantly impinge upon our society result from our planet's need to rid itself of the heat that is constantly generated in the interior by the decay of radioactive elements. As in the Earth's early history, this is carried towards the surface by convection currents within the mantle. These currents in turn constitute the engines that drive the great, rocky plates across the surface of the planet, and underpin the concept of *plate tectonics,* which geophysicists use to provide a framework for how the Earth operates geologically.

The relative movements of the plates themselves, which comprise the crust and the uppermost rigid part of the mantle (together known as the *lithosphere*), are in turn directly related to the principal geological hazards—earthquakes and volcanoes, which are concentrated primarily along plate margins. Here a number of interactions are possible. Two plates may scrape jerkily past one another, accumulating strain and releasing it periodically through destructive earthquakes. Examples of such *conservative* plate margins include the quake-prone San Andreas Fault that separates western California from the rest of the United States and Turkey's North Anatolian Fault, whose latest movement triggered a major earthquake in 1999. Alternatively, two plates may collide head on. If they both carry continents built from low-density granite rock, as with the Indian Ocean and Eurasian plates, then the result of collision is the growth of a high mountain range—in this case the Himalayas—and at the same time the generation of major quakes such as that which obliterated the Indian city of Bhuj in January 2001. On the other hand, if an oceanic plate made of dense basalt hits a low-density continental plate then the former will plunge underneath, pushing back into the hot, convecting mantle. As one plate thrusts itself beneath the other (a process known as *subduction*) so large earthquakes are generated. Subduction is going on all around the Pacific Rim, ensuring high levels of seismic activity in Alaska, Japan, Taiwan, the Philippines, Chile, and elsewhere in the circum-Pacific region. This type of *destructive* plate margin—so called because one of the two colliding plates is destroyed—also hosts large numbers of active volcanoes. Although the mechanics of magma formation in such regions is sometimes complex, it is ultimately a result of the subduction process and owes much to the partial melting of the subducting plate as it is pushed down into ever hotter levels in the mantle. Fresh magma formed in this way rises as a result of its low density relative to the surrounding rocks, and blasts its way through the surface at volcanoes that are typically explosive and particularly hazardous. Strings of literally hundreds of active and dormant volcanoes circle the Pacific, making up the legendary *Ring of Fire*, while others sit above subduction zones in the Caribbean and Indonesia. Virtually all large, lethal eruptions occur in these areas, and recent volcanic disasters have occurred at Pinatubo (Philippines) in 1991, Rabaul (Papua New Guinea) in 1994, and Montserrat (Lesser Antilles, Caribbean) from 1995 until the time of writing.

To compensate for the consumption of some plate material, new rock must be created to take its place. This

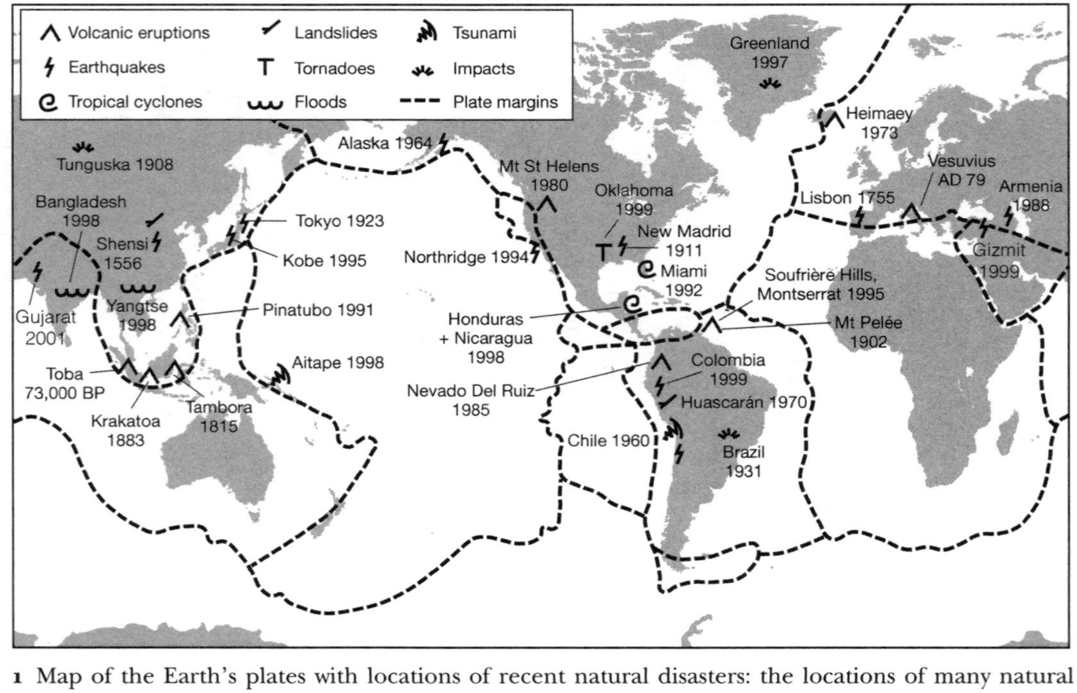

1 Map of the Earth's plates with locations of recent natural disasters: the locations of many natural disasters coincide with the plate margins

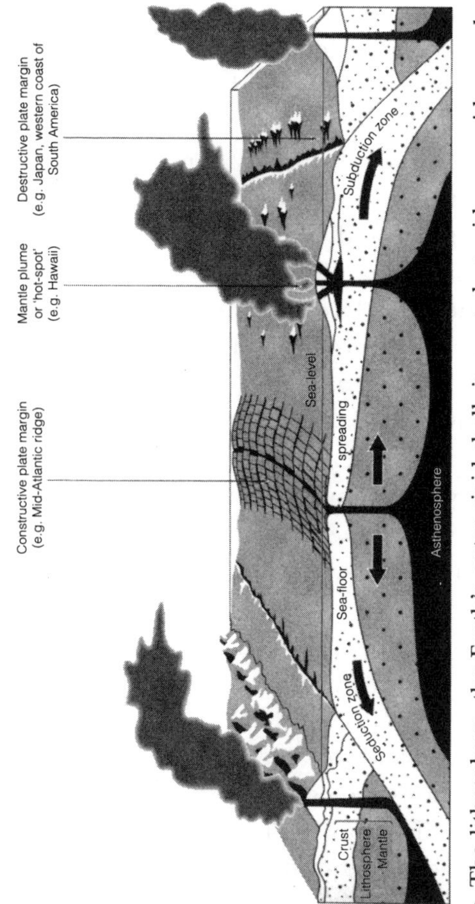

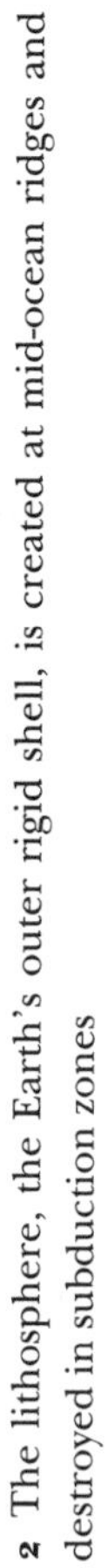

2 The lithosphere, the Earth's outer rigid shell, is created at mid-ocean ridges and destroyed in subduction zones

happens at so-called *constructive* plate margins, along which fresh magma rises from the mantle, solidifies, and pushes the plates on either side apart. This occurs beneath the oceans along a 40,000-kilometre long network of linear topographic highs known as the *Mid-Ocean Ridge system*, where newly created lithosphere exactly balances that which is lost back into the mantle at destructive margins. A major part of the Mid-Ocean Ridge system runs down the middle of the Atlantic Ocean, bisecting Iceland, and separating the Eurasian and African plates in the east from the North and South American plates in the west. Here too there are both volcanoes and earthquakes, but the former tend to involve relatively mild eruptions and the latter are small. Driven by the mantle convection currents beneath, the plates waltz endlessly across the surface of the Earth, at about the same rate as fingernails grow, constantly modifying the appearance of our planet and ensuring that, given time, everywhere gets its fair share of earthquakes and volcanic eruptions.

Hazardous Earth

While earthquakes and volcanic eruptions are linked to how our planet functions geologically, other geophysical hazards are more dependent upon processes that operate in the Earth's atmosphere. Rather than the heat from the interior, our planet's weather machine is driven by energy from the Sun. Our nearest star is the ultimate instigator—aided by the Earth's rotation and the constant exchange of energy and water with the oceans—of the tropical cyclones and floods that exact an enormous toll on life and property, particularly in developing countries. Still other lethal natural phenomena have a composite origin and are less easy to pigeonhole. The giant sea waves known as *tsunamis* (or sometimes incorrectly as 'tidal waves'), for example, can be formed in a number of different ways; most commonly by submarine earthquakes, but also by landslides into the ocean and by eruptions of coastal and island volcanoes. Similarly, many landslides result from a collusion between geology and meteorology, with torrential rainfall destabilizing already weak slopes. Although there remains an enormous amount to learn about natural hazards, their causes and characteristics, our current level of knowledge is truly encyclopedic—and if so desired you can indeed consult weighty and authoritative tomes focused entirely on specific hazards. Here, as a taster, my intention is to gallop you through the principal features of the major natural hazards at a pace which I hope is not too great, before placing their current and future impact on our society in some perspective.

At any single point and at any one time the Earth and its enclosing atmospheric envelope give the impression of being mundanely stable and benign. This is, however, an entirely misleading notion, with something like 1,400 earthquakes rocking the planet every day and a volcano erupting every week. Each year, the tropics are battered by up to 40 hurricanes, typhoons, and cyclones, while floods and landslides occur everywhere in numbers too great to keep track of.

In terms of the number of people affected—at least 100

million people a year—floods undoubtedly constitute the greatest of all natural hazards, a situation that is likely to continue given a future of rising sea levels and more extreme precipitation. River floods are respecters of neither wealth nor status, and both developed and developing countries have been severely afflicted in recent years, across every continent. Wherever rain is unusually torrential or persistent, it will not be long before river catchments fail to contain surface run-off and start to expand across their flood plains and beyond. In fact, the intensity of rainfall can be quite astonishing, with, in 1970, nearly 4 centimetres of rain falling in just 60 *seconds* on the French Caribbean island of Guadeloupe—a world record. On another French island, Réunion, in the Indian Ocean, a passing cyclone dropped close to 4 *metres* of rain during a single 24-hour period in March 1952. As flood plains all over the world become more crowded, the loss of life and damage to property caused by swollen rivers has increased dramatically. In the spring of 1993, the Mississippi and Missouri rivers burst their banks, inundating nine Midwest states, destroying 50,000 homes and leaving damage totalling 20 billion US$. Massive floods occurred in many parts of the UK in autumn 2000 as rain fell with a ferocity not seen for over 300 years. River flooding continues to pose a major threat in China, and has been responsible for over 5 million deaths over the last 150 years. Bangladesh has it even worse, with the country often finding two-thirds of its land area under water as a result either of floodwaters pouring down the great Ganges river system or of cyclone-related storm surges pouring inland from the Bay of Bengal. Coastal

3 A tornado. Tornadoes contain the strongest winds on Earth, sometimes in excess of 500 kilometres an hour

flooding due to storms probably takes more lives than any other natural hazard, with an estimated 300,000 losing their lives in Bangladesh in 1970 and 15,000 at Orissa, northeast India, in 1999.

Partly through their effectiveness at spawning floods, but also through the enormous wind speeds achieved, storms constitute one of the most destructive of all natural hazards. Furthermore, because they are particularly common in some of the world's most affluent regions, they are responsible for some of the most costly natural disasters of all time. Every year, the Caribbean, the Gulf and southern states of the USA, and Japan are struck by tropical storms, while the UK and continental Europe suffer increasingly from severe and damaging winter storms. In 1992, Hurricane Andrew virtually obliterated southern Miami in one of the costliest natural disasters in US history, resulting in losses of 32 billion US$. This epic storm brought to bear on the city wind speeds of up to 300 kilometres per second, leaving 300,000 buildings damaged or destroyed and 150,000 homeless. Destructive windstorms are not only confined to the tropics, and hurricane-force winds also accompany low-pressure weather systems at mid-latitudes. Many residents of southern England will remember the Great Storm of October 1987 that felled millions of trees with winds whose average speeds were clocked at just below hurricane force. More recently, in 1999, France suffered a similar ordeal as winter storm Lothar blasted its way across the north of the country. Across the ocean, the US Midwest braces itself every year for a savage onslaught from tornadoes: rotating maelstroms of solid wind

4 Earthquake damaged buildings. As many as 100,000 people may have died in the earthquake in Gujarat, India (January 2001)

that form during thunderstorms in the contact zone between cold, dry air from the north and warm, moist air from the tropics. No man-made structures that suffer a direct hit can withstand the average wind speeds of up to 500 kilometres an hour, and damage along a tornado track is usually total. Although rarely as lethal as hurricanes, in just a few days in April 1974 almost 150 tornadoes claimed over 300 lives in Kentucky, Tennessee, Alabama, and adjacent states.

Of the so-called geological hazards—earthquakes, volcanic eruptions, and landslides—there is no question that earthquakes are by far the most devastating. Every year about 3,000 quakes reach magnitude 6 on the well-known Richter Scale, which is large enough to cause significant damage and loss of life, particularly when they strike poorly constructed and ill-prepared population centres in developing countries. As previously mentioned, most large earthquakes are confined to distinct zones that coincide with the margins of plates. In recent years, sudden movements of California's San Andreas Fault have generated large earthquakes in San Francisco (1989) and southern California (1994), the latter causing damage totalling 35 billion US$—the costliest natural disaster in US history. Just a year later, a magnitude 7.2 quake at the western margin of the Pacific plate devastated the Japanese city of Kobe, killing 6,000 and engendering economic losses totalling a staggering 200 billion US$—the most expensive natural disaster of all time. Four years after Kobe, the North Anatolian Fault slipped just to the east of Istanbul, generating a severe quake that flattened the town of Izmit and neighbouring settlements and took over 17,000

5 In 1998 17-metre-high tsunamis devastated the north coast of Papua New Guinea, taking close to 3,000 lives

lives. Large earthquakes can also occur, however, at locations remote from plate margins, and have been known in northern Europe and the eastern USA, which are not regions of high seismic risk. The last such *intraplate* quake devastated the Bhuj region of India's Gujarat state in January 2001, completely destroying 400,000 buildings and killing perhaps as many as 100,000 people. There is a truism uttered by earthquake engineers: *it is buildings not earthquakes that kill people.* Without question this is the case, and both damage to property and loss of life could be drastically reduced if appropriate building codes were both applied and enforced. Earthquakes, however, also prove lethal through the triggering of landslides as a result of ground shaking, and by the formation of tsunamis. The latter are generated when a quake instantaneously jerks upwards—perhaps by just a metre or so—a large area of the seabed, causing the displaced water above to hurtle outwards as a series of waves. When these enter shallow water they build in height—sometimes to 20 metres or more—and crash into coastal zones with extreme force. In 1998, Sissano and neighbouring villages on the north coast of Papua New Guinea were wiped out and 3,000 of their inhabitants drowned or battered to death by a 17-metre-high tsunamis that struck within minutes of an offshore earthquake.

Estimates of the number of active volcanoes vary, but there are at least 1,500 and possibly over 3,000. Every year around 50 volcanoes erupt, some of which—like Kilauea on Hawaii or Stromboli in Italy—are almost constantly active. Others, however, may have been quiet for centuries or in some cases millennia and these tend to be the most destructive. The most violent volcanoes occur at destructive plate margins, where one plate is consuming another. Their outbursts rarely produce quiet flows of red lava and are more likely to blast enormous columns of ash and debris 20 kilometres or more into the atmosphere. Carried by the wind over huge areas, volcanic ash can be extremely disruptive, making travel difficult, damaging crops, poisoning livestock, and contaminating water supplies. Just 30 centimetres or so of wet ash is sufficient to cause roofs to collapse while the fine component of dry ash can cause respiratory problems and illnesses such as silicosis. Close to an erupting volcano the depth of accumulated ash can total several metres, sufficient to bury single-storey structures. This was the fate of much of the town of Rabaul on the island of New Britain (Papua New Guinea), during the 1994 eruptions of its twin volcanoes Vulcan and Tavurvur. For years following the 1991 eruption of Pinatubo in the Philippines, thick deposits of volcanic debris provided a source for mudflows whenever a tropical cyclone passed overhead and dumped its load of rain. Almost a decade later, mud pouring off the volcano was still clogging rivers, inundating towns and agricultural land, and damaging fisheries and coral reefs. Somewhat surprisingly, mudflows also constitute one of the biggest killers at active volcanoes. In 1985 a small eruption through the ice and snow fields of Columbia's Nevado del Ruiz volcano unleashed a torrent of mud out of all proportion to the size of the eruption, which poured down the valleys draining the volcano and buried the town of Armero and 23,000 of its inhabitants.

6 Ruins of St Pierre (Martinique) after 1902 eruption: only two inhabitants of St Pierre survived the onslaught of the Mont Pelée volcano

Even scarier and more destructive than volcanic mudflows are *pyroclastic flows* or *glowing avalanches.* These hurricane-force blasts of incandescent gas, molten lava fragments, and blocks and boulders sometimes as large as houses have the power to obliterate everything in their paths. In 1902, in the worst volcanic disaster of the twentieth century, pyroclastic flows from the Mont Pelée volcano on the Caribbean island of Martinique annihilated the town of St Pierre as effectively as a nuclear bomb, within a few minutes leaving only two survivors out of a population of 29,000. The threat from volcanoes does not end there: chunks of rock collapsing from their flanks can trigger huge tsunamis, while noxious fumes can and have locally killed thousands and their livestock. Volcanic gases carried into the stratosphere, and from there around the planet, have modified the climate and led to miserable weather, crop failures, and health problems half a world away. On the grandest scale, volcanic super-eruptions have the potential to affect us all, through plunging the planet into a frigid *volcanic winter* and devastating harvests worldwide.

Of all geological hazards, landslides are perhaps the most underestimated, probably because they are often triggered by some other hazard, such as an earthquake or deluge, and the resulting damage and loss of life is therefore subsumed within the tally of the primary event. Nevertheless, landslides can be highly destructive, both in isolation and in numbers. In 1556, a huge earthquake struck the Chinese province of Shensi, shaking the ground so vigorously that the roofs of countless cave dwellings collapsed, incarcerating (according to Imperial records) over 800,000 people. In 1970, another quake caused the entire peak of the Nevados Huascaran mountain in the Peruvian Andes to fall on the towns below, wiping out 18,000 people in just four minutes and erasing all signs of their existence from the face of the Earth. Heavy rainfall too can be particularly effective at triggering landslides, and when in 1998 Hurricane Mitch dumped over 30 centimetres of rain on Central America, it mobilized over a million landslides in Honduras alone, blocking roads, burying farmland, and destroying communities.

The final—and perhaps greatest—threat to life and limb comes not from within the Earth but from without. Although the near constant bombardment of our planet by large chunks of space debris ended billennia ago, the threat from asteroids and comets remains real and is treated increasingly seriously. Even as I write, the UK government has announced funding for a new research centre dedicated to the study of the impact threat and its consequences. Recent estimates suggest that around a thousand asteroids with diameters of 1 kilometre or more have orbits around the Sun that cross the Earth's, making collision possible at some point in the future: 1 kilometre is the impactor diameter threshold for initiating a *cosmic winter*, due to dust lifted into the stratosphere blocking out solar radiation, for wiping out a quarter or so of the human population, and for causing general mayhem worldwide. The revival of interest in the impact threat has arisen as a result of two important scientific events during the last decade: first, the identification of a large impact crater at Chicxulub, off Mexico's Yucatan Peninsula, which has now been established as the 'smoking gun' responsible, ultimately, for global genocide at the end of the Cretaceous period: second, the eye-opening collisions in 1994 of the fragments of Comet Shoemaker-Levy with Jupiter. Images flashed around the world of resulting impact scars larger than our own planet were disconcerting to say the least and begged the question in many quarters—what if that were the Earth?

Natural hazards and us

If you were not already aware of the scale of the everyday threat from nature then I hope, by now, to have engendered a healthy respect for the destructive potential of the hazards that many of our fellow inhabitants of planet Earth have to face almost on a daily basis. The reinsurance company Munich Re., who, for obvious reasons, have a considerable interest in this sort of thing, estimate that up to 15 million people were killed by natural hazards in the last millennium, and over 3.5 million in the last century alone. At the end of the second millennium AD, the cost to the global economy reached unprecedented levels, and in 1999 storms and floods in Europe, India, and South East Asia, together with severe earthquakes in Turkey and Taiwan and devastating landslides in Venezuela, contributed to a death toll of 75,000 and economic losses totalling 100 billion US$.

The last three decades of the twentieth century each saw a billion or so people suffer due to natural disasters. Unhappily, there is little sign that hazard impacts on society have diminished as a consequence of improvements in forecasting and hazard mitigation, and the outcome of the battle against nature's dark side remains far from a foregone conclusion. While we now know far more about natural hazards, the mechanisms that drive them, and their sometimes awful consequences, any benefits accruing from this knowledge have been at least partly negated by the increased vulnerability of large sections of the Earth's population. This has arisen primarily as a result of the rapid rise in the size of the world's population, which doubled between 1960 and 2000. The bulk of this rise has occurred in poor developing countries, many of which are particularly susceptible to a whole spectrum of natural hazards. Furthermore, the struggle for

Lebensraum has ensured that marginal land, such as steep hillsides, flood plains, and coastal zones, has become increasingly utilized for farming and habitation. Such terrains are clearly high risk and can expect to succumb on a more frequent basis to, respectively, landsliding, flooding, storm surges, and tsunamis.

Another major factor in raising vulnerability in recent years has been the move towards urbanization in the most hazard-prone regions of the developing world. Within just a few years, and for the first time ever, more people will live in urban environments than in the countryside, many crammed into poorly sited and badly constructed *megacities* with populations in excess of 8 million people. Forty years ago New York and London topped the league table of cities, with populations, respectively, of 12 and 8.7 million. In 2015, however, cities such as Mumbai (formerly Bombay, India), Dhaka (Bangladesh), Karachi (Pakistan), and Mexico City will be firmly ensconced in the top ten (Table 1): gigantic sprawling agglomerations of humanity with populations approaching or exceeding 20 million, and extremely vulnerable to storm, flood, and quake. A staggering 96 per cent of all deaths arising from natural hazards and environmental degradation occur in developing countries and there is currently no prospect of this falling. Indeed, the picture looks as if it might well deteriorate even further. With so many people shoehorned into ramshackle and dangerously exposed cities it can only be a matter of time before we see the first of a series of true *mega disasters*, with death tolls exceeding one million.

Table 1 The predicted ten most populous cities in the world in 2015

City	*Population*
1. Tokyo	28,900,000
2. Mumbai	26,900,000
3. Lagos	24,600,000
4. Sao Paulo	20,300,000
5. Dhaka	19,500,000
6. Karachi	19,400,000
7. Mexico City	19,200,000
8. Shanghai	18,000,000
9. New York	17,600,000
10. Calcutta	17,300,000

Source: Munich Reinsurance 1999.

The picture I have painted is certainly bleak, but the reality may be even worse. Future rises in population and vulnerability will take place against a background of dramatic climate change, the like of which the planet has not experienced for maybe 10,000 years. The jury remains out on the precise hazard implications of the rapid warming expected over the next hundred years, but rises in sea level that may exceed 80 centimetres are forecast in the most recent (2001) report of the IPCC (Intergovernmental Panel on Climate Change). This will certainly increase the incidence and impact of storm surges and tsunamis and—in places—raise the level of coastal erosion. Other consequences of a temperature rise that could reach 6 degrees Celsius by the end of the century may include more extreme meteorological events such as hurricanes, tornadoes, and floods, greater numbers of landslides in mountainous terrain, and, eventually, even more volcanic eruptions (see next chapter).

So is the world as we know it about to end and, if so, how? A century from now will we be gasping for water in an increasingly roasting world or huddling around a few burning sticks, struggling to keep at bay the bitter cold of a cosmic winter? In the next chapter I will delve a little further into the possibilities.

Facts to contemplate

- During its earliest history, the Earth was covered in a magma ocean with temperatures—at 5,000 degrees Celsius—comparable with the surfaces of some of the cooler stars.
- Our planet's great tectonic plates move at about the same rate that our fingernails grow.
- Around 1,400 earthquakes rock the planet every day.
- There may be 3,000 or more active or potentially active volcanoes, about 50 of which erupt every year.
- The tropics are battered every year by 40 or more hurricanes, typhoons, and cyclones.
- In 1556 a single earthquake in the Shensi province of China is estimated to have killed 800,000 people.

- At least 15 million deaths in the last millennium are attributed to natural hazards.
- 96 per cent of all deaths from natural hazards and environmental degradation now occur in developing countries.

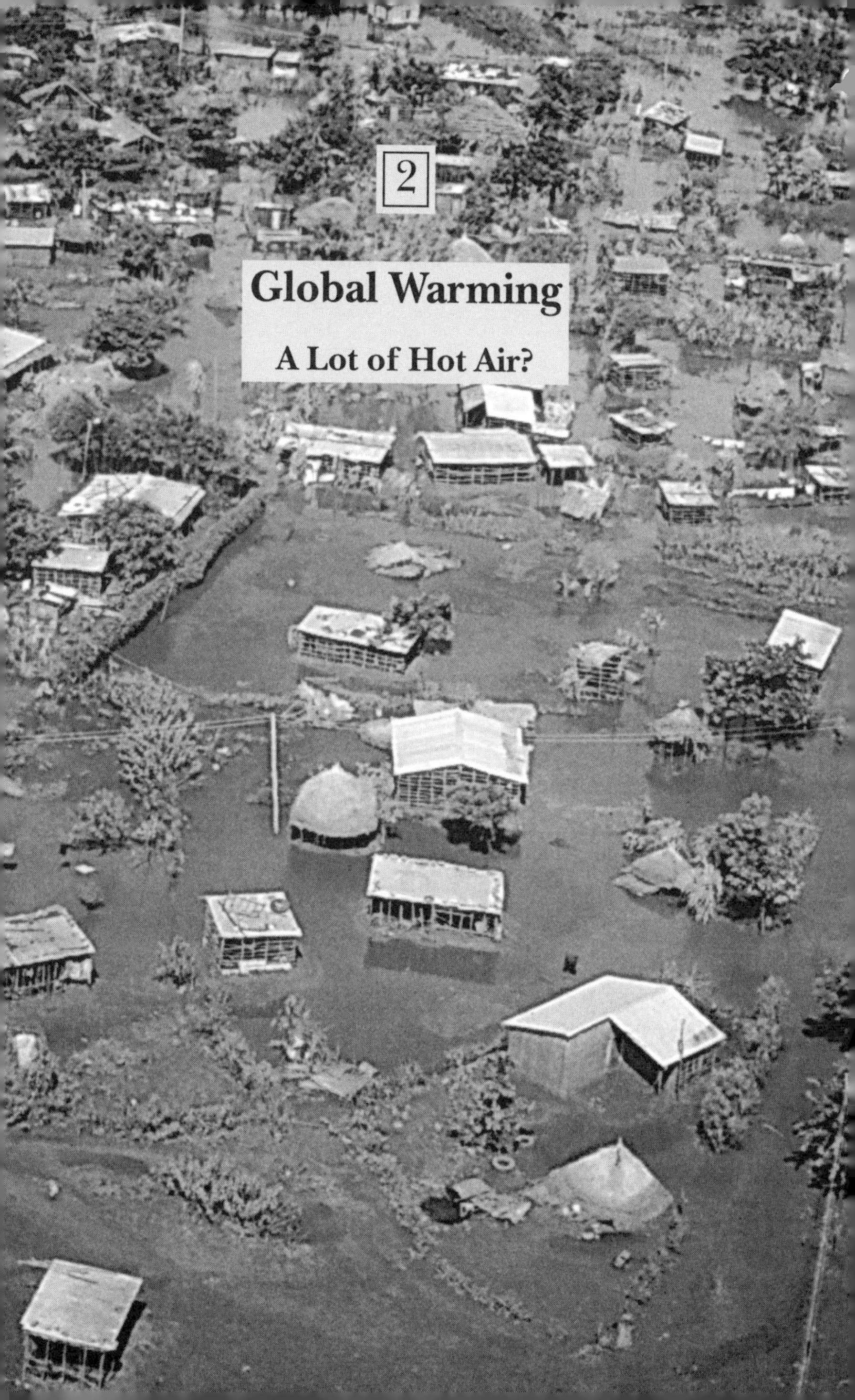

2

Global Warming

A Lot of Hot Air?

Debate—what debate?

Global warming is about much more than hotter summers, winter floods, and farting cows. There is absolutely no question that the Earth is warming up fast, and few climate scientists would argue with this. The dispute lies in whether or not the warming we are now experiencing simply reflects a natural turnabout in the recent global temperature trend or results from the polluting impact of human activities since the industrial revolution really began to take hold. What I find extraordinarily irresponsible is that this dispute continues to be presented, at least in some circles, as a battle between two similarly sized and equally convincing schools of scientific thought, when in fact this is far from the case. Forecasting climate change is extremely difficult, which explains why models for future temperature rise and sea-level change are constantly undergoing revision, but the evidence is now irrefutable: human activities *are* driving the current period of planetary warming.

Notwithstanding a few maverick scientists, oil company representatives, and the president of the world's greatest polluter, the overwhelming consensus amongst those who have a grasp of the facts is that without a reduction in greenhouse gas emissions things are going to get very bad indeed. Amazingly, this prospect is still played down and intentionally hidden behind a veil of obfuscation by some, most recently by the—in my opinion—self-deluded Danish statistician, Blom Lomburg. In his recently published and widely savaged book, *The Skeptical Environmentalist*, Lomburg denigrates global warming and its future impact, while at the same time, through highly selective references to scientific research, coming to the conclusion that all is right with the world. Just in case you have come across this work and been lulled by its friendly, do-nothing message into a false sense of security, let me bring you back to reality, if I may, with a few pertinent facts.

During the past 70 years, the Earth has been hotter than at any other time in the last millennium, and the warming has accelerated dramatically in just the past few decades. No doubt everyone has at least one older relative who is constantly harking back—through a rose-tinted haze—to a time when summers were hotter and the skies bluer. Meteorological records show, however, that this is simply a case of selective memory, and in fact 15 of the hottest 16 years on record have occurred since 1980, with the late 1990s seeing the warmest years of all across the planet as a whole. The Earth is now warmer than it has been for over 90 per cent of its 4.6 billion year history, and by the end of the twenty-first century our planet may see higher temperatures than at any time for the last 150,000 years.

The rising temperature trend we are seeing now is not simply a climatic blip or hiccup, nor can it be explained entirely, as some would still have it, by variations in the output of the Sun, although this clearly does have a significant effect on the climate. Rather, it is a consequence of two centuries of pollution, which is now enclosing the Earth in an insulating blanket of carbon dioxide, methane, nitrous oxide, and other greenhouse gases. Since the late eighteenth century our race has been engaged in a gigantic planetary trial, the final outcome of which we can still only guess at. Unfortunately for us the experiment has now entered a runaway phase, which, due to its inherent inertia, we cannot stop but only slow down. Even if we were to stabilize greenhouse gas emissions today, both temperatures and sea levels would continue to rise for many hundreds of years. The big question of our time is—do we have the resolve to do even this or will we run from the problem and let the devil take the hindmost? Let's head for the laboratory and see how things are progressing.

The great global warming experiment

We know from studies of polar ice cores that before the hiss of steam and grinding of metal on metal that heralded the arrival of the industrial world, the concentrations of greenhouse gases in the atmosphere had been pretty much constant since the glaciers retreated at the end of the last Ice Age. Since pre-industrial times, however, carbon dioxide levels in the atmosphere have risen by 30 per cent, alongside sharp increases in other greenhouse gases, in particular methane and nitrous oxide. Atmospheric concentrations of carbon dioxide levels are now higher than they have been for at least 420,000 years and may not have been exceeded during the past 20 *million* years. The rate of increase in the gas has also been quite unprecedented, and was greater in the last hundred years than at any time in at least the previous 20,000. Being concoctions of the twentieth century, other polluting gases such as chlorofluorocarbons and hydrofluorocarbons were not even present in the atmosphere a couple of centuries ago. As these gases have accumulated in the Earth's atmosphere so they have, quite literally, caused it to act in the manner of a greenhouse, allowing heat from the Sun in but hindering its escape back out into space. In fact, our atmosphere has operated in this way for billions of years, moderating temperature swings and extremes, but our pollution is now strongly enhancing this greenhouse effect, with the result that the Earth has been progressively warming up for most of the last hundred years.

Because the climate machine is so complex, however, no single influence can be taken in isolation and many other factors affect global temperatures. Not least of these is the output of the Sun, which is also variable over time, and which must be taken into account before allocating a rising temperature trend purely to the accumulation of man-made greenhouse gases. The Sun follows a regular 11-year pattern of activity, known as the *sunspot cycle*, during which time its output varies by about 0.1 per cent. Solar output also changes over longer periods, ranging from hundreds to tens of thousands of years, and these can play a significant role in cooling or warming the planet and—in recent centuries—in modifying or masking the effect of anthropogenically

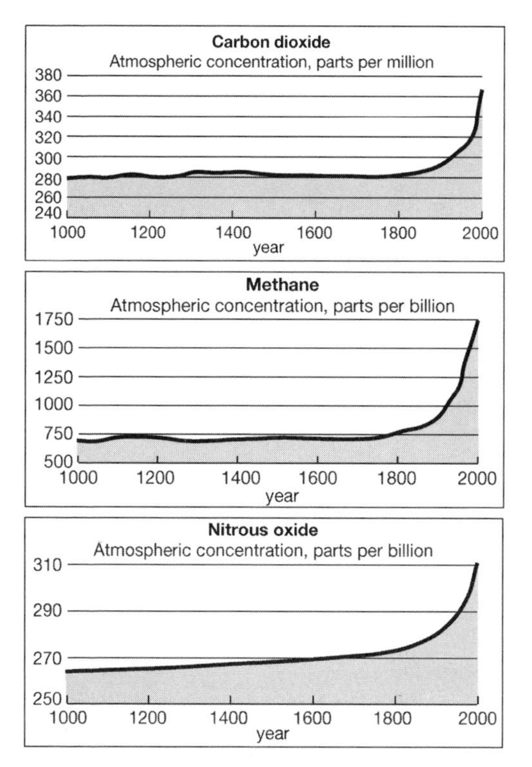

7 Concentrations of greenhouse gases have risen dramatically since the industrial revolution

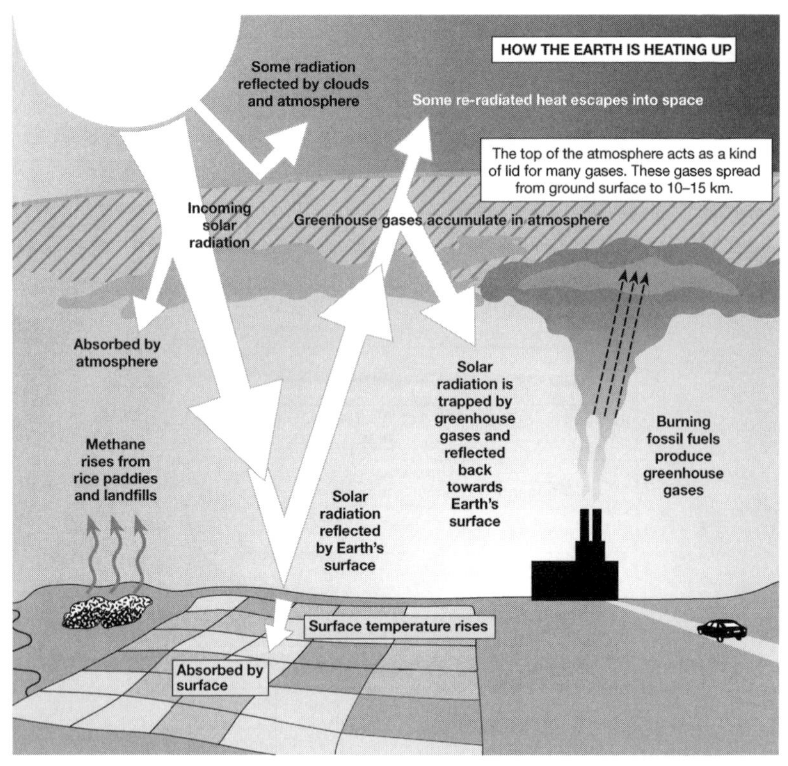

8 Greenhouse gases accumulating in the atmosphere and forming an insulating blanket that is rapidly warming the planet

derived gases. As I will address in more detail in Chapter 4, volcanic eruptions can also have a significant effect on the Earth's climate. Although the detailed picture is somewhat more complex, large explosive eruptions inject massive volumes of sulphur dioxide and other sulphur gases into the stratosphere, which have a broadly cooling effect through reducing the level of solar radiation reaching the Earth's surface. Significant, if short-lived, reductions in global temperatures followed the eruptions of both El Chichón (Mexico) in 1982 and Pinatubo (Philippines) in 1991. Sometimes volcanoes and the Sun combine to bring about longer-lasting episodes of climate change. For example, a combination of reduced solar output and elevated volcanic activity has been implicated in the medieval cold snap known by climate scientists as the *Little Ice Age*. This lasted from about 1450 AD to perhaps the end of the nineteenth century and saw frost fairs on the Thames and bitter winters in many parts of the world.

Attempting to pin down the true variation in global temperatures over the past thousand years is difficult, not least because records prior to the last couple of hundred years are far from reliable. A further complication arises from the fact that while one part of the world might be heating up, another might be cooling down. One argument that is still used by opponents of anthropogenic warming is that the world underwent a pronounced cooling between 1946 and 1975, thereby invalidating the idea that elevated levels of greenhouse gases must automatically result in global warming. More detailed examination of the record for this period reveals, however, that although much of the northern

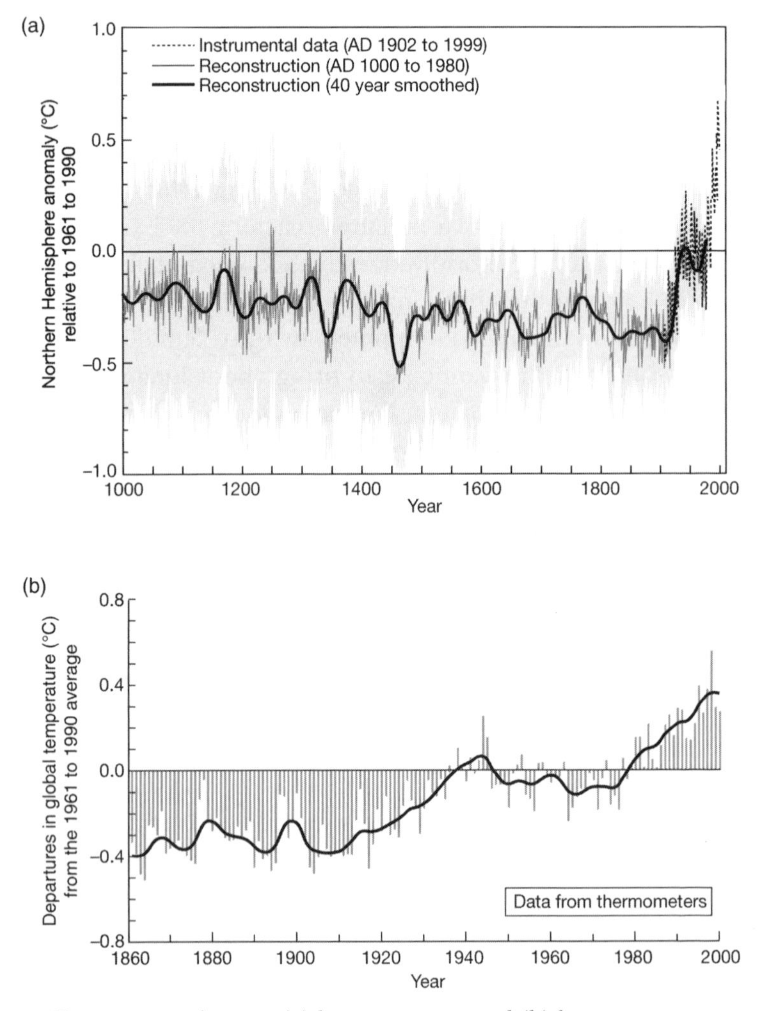

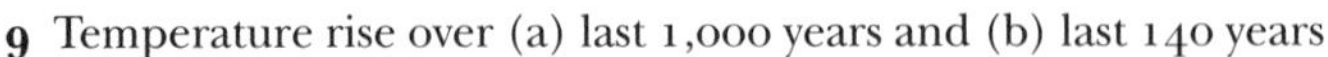
9 Temperature rise over (a) last 1,000 years and (b) last 140 years

hemisphere cooled noticeably, the reverse was the case in the southern hemisphere, which warmed appreciably. In fact, although there was a small overall temperature fall at this time, this is now being attributed to a masking of the warming effect by sulphur gases emitted by volcanic eruptions and by heavy industries at the time unfettered by clean air laws. Notwithstanding this blip, instrumental records show that temperatures have been following an inexorably upward path since such records began in 1861. The record also shows that the 1990s was the warmest decade since the mid-nineteenth century and 1998 the warmest year on record.

If our great experiment was designed specifically to heat up the planet then based upon the results to date it seems that we can pat one another on the back at a job well done and sit back and relax as the experiment grinds away of its own accord, racking up the heat and clocking up an ever increasing list of unexpected consequences. But of course, this was not the intention of the experiment at all. Indeed, it is only in recent decades that the polluting effect of human activities on the global environment has been thought of in these terms. The great experiment has never been anything other than a side effect of our race's constant thirst for more; more growth, more goods, more wealth. Now that it has become apparent that we have been messing, admittedly involuntarily, with the natural functioning of the Earth we have no choice but to close the experiment down. Continuing political procrastinations and the muddying of the scientific waters by vested interest groups antagonistic to proposals to mitigate global warming has ensured that the

Kyoto Protocol is well and truly dead. Even had it been ratified and achieved its goal of a 5.2 per cent reduction (below 1990 levels) in greenhouse gas emissions by 2008–12, this would have had little impact on the rate of continued warming. For any real impact to be made on the rate of warming, reductions in emissions of the order of 60 per cent are needed *now*.

Rather like trying to turn around a supertanker, the enormous inertia that has already built up in the system would still mean that, even if we came to our senses and made dramatic cuts in emissions within the next few years, both temperatures and sea levels would continue to rise for centuries to come. It seems inevitable, therefore, that we are going to face dramatic changes to our environment—some for the better, but most not. What is certain is that our children and their descendants are going to find the Earth a very different place.

Hothouse Earth

The world of 2100 AD will not only be far warmer but will also be characterized by extremes of weather that will ensure, at the very least, a far more uncomfortable life for billions. Already, the wildly fluctuating weather patterns that are held by many to be a consequence of global warming, combined with increased vulnerability in the developing world, are leading to a dramatic rise in the numbers of meteorological disasters. In its 2001 World Disasters Report, the International Federation of Red Cross and Red Crescent Societies reveals that the annual numbers of disasters due to storms, floods, landslides, and droughts have climbed from around 200 before 1996 to almost 400 in the first year of the millennium. Few think that the situation will get better and the chances are that things will get progressively worse.

Increasingly, those occupying low-lying coastal regions will be hit by rising sea levels and heavier rainfall that will mean that lethal floods become the norm rather than the exception. In contrast, more and more people will starve as annual rains fail year after year and huge regions of Africa and Asia fall within the grip of drought and consequent famine. It also looks as if the Earth will become a windier place, with warmer seas triggering more and bigger storms, particularly in the tropics. I will return to the manifold hazard implications of global warming later, but let's look now at the latest predictions for temperature rise over the next 100 years. After all, this is the critical element that will drive the huge changes to our environment in this century and beyond.

Earlier this year the Intergovernmental Panel on Climate Change, or IPCC, published its Third Assessment Report on global warming; three massive tomes totalling over 2,600 pages. The IPCC was established in 1988 by the UN Environmental Programme and the World Meteorological Organization, with a remit to provide an authoritative consensus of scientific opinion on climate change using the best available expertise. The important word here is *consensus*. Over 1,000 scientists were involved in either the writing of the report or the reviewing of its content, leaving little doubt of its validity except in the minds of the irrationally sceptical, the eternally optimistic, or the downright Machiavellian. If the content of the third IPCC report could be summed up in a few words, they would probably be 'Did we say in our 1995 report that things would be bad? Well, we were wrong. They are going to be much worse than that.'

Let's look at what the panel says about rising temperatures. Over the course of the last century, global temperatures rose by 0.6 degrees Celsius. By 2100, the IPCC worst case scenario predicts that temperatures will be almost 6 degrees Celsius higher than they are now, and even the average prediction would see us roasting as a consequence of a 4 degree Celsius rise. If this does not sound much, consider that just 4 or 5 degrees Celsius mean the difference between full Ice Age conditions and our current climate. The transition between the two involved huge changes in the Earth's environment, not only in the climate and weather but also in vegetation and animal life.

There is every reason to expect that as the post-glacial temperature rise doubles again we will experience equally dramatic changes. This time, however, there are two important differences. First, the Earth has to feed, clothe, and support 6 billion or more souls, rather than a few million, and secondly today's comparable temperature rise is taking place over the course of just a hundred years rather than thousands. Many of the consequences of such a dramatic rise in global temperatures are obvious, but others less so. The polar regions and mountainous areas with permanent snow and ice are already suffering, and warming will continue to exact a severe toll here. Over the last 100 years there has been a massive retreat of mountain glaciers all over the world, while since the 1950s the Arctic ice has started to thin dramatically with the result that the North Pole was ice free last summer. Furthermore, the extent of Arctic sea ice in spring and summer is 10–15 per cent smaller than it was 40 years ago, while ice on lakes and rivers at higher altitudes in the northern hemisphere now melts in spring two weeks earlier than a century ago. Northern hemisphere spring snow cover is already 10 per cent down on the 1966–86 mean and IPCC predictions suggest that polar and mountainous regions of the hemisphere could be 8 degrees Celsius warmer by 2100. In 1999, the concentration of carbon dioxide in the atmosphere was 367 parts per million (ppm). Even if, at some future time, we managed to stabilize the concentration at 450 ppm, temperatures would continue to rise, albeit more slowly, beyond the year 2300.

Dramatically increasing the rate of melting of snow and ice means rising sea levels: tide gauge data indicate that global sea levels rose by between 10 and 20 centimetres during the twentieth century, and this rise is expected to escalate drastically in the coming hundred years, with sea levels predicted to be 40 centimetres and perhaps over 80 centimetres higher by 2100. Most of the recent and predicted rise comes from the thermal expansion of the oceans as they warm up or by the addition of water from the rapidly melting mountain glaciers. Failure to cut back on greenhouse emissions, however, may lead in future to catastrophic melting of the Greenland and Antarctic ice sheets, resulting in terrible consequences for coastal areas. Worst case scenarios in the IPCC report

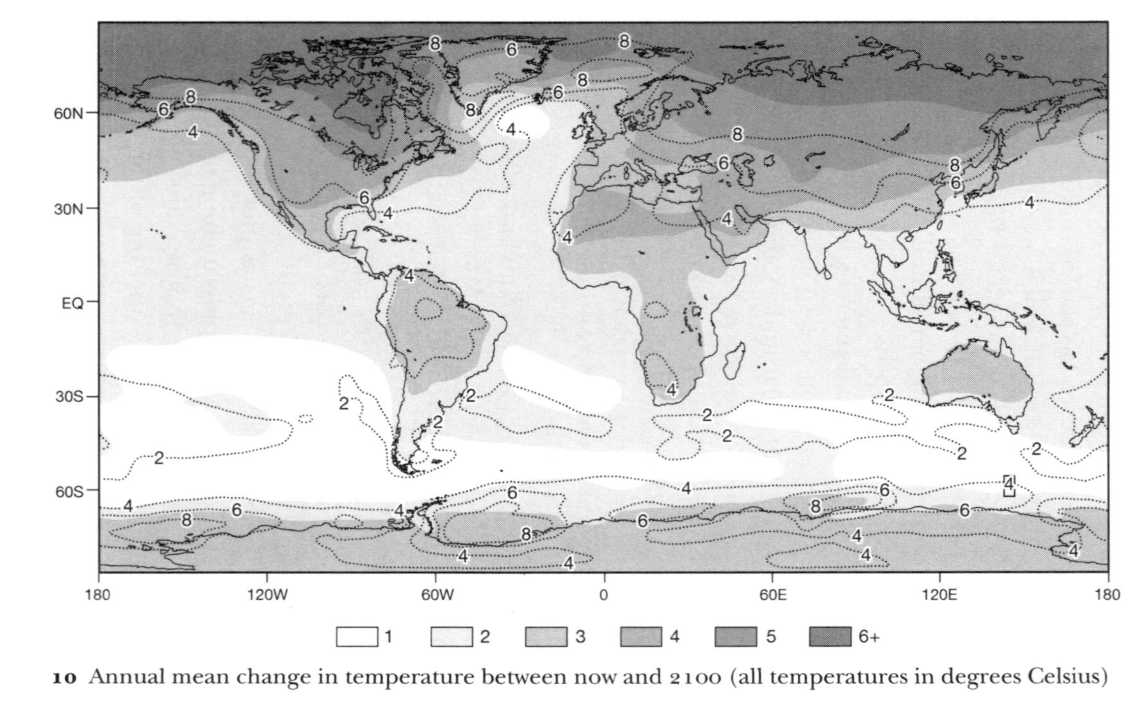

10 Annual mean change in temperature between now and 2100 (all temperatures in degrees Celsius)

forecast the near elimination of the Greenland Ice Sheet, generating a 6-metre rise in sea level by the year 3000. The great West Antarctic Ice Sheet appears at present to be more stable, but severe warming over the next few millennia could result in its permanent disintegration and loss. Should the Greenland Ice Sheet melt fully, then virtually all the world's major coastal cities will find themselves under water. Even without this, however, the effects of rising sea level in the next hundred years will be devastating for low-lying countries. For example, a 1-metre rise would see the Maldives in the Indian Ocean under water, while a combination of rising

11 Devastating floods, such as those which affected Mozambique in 2000, are likely to become more frequent as the Earth continues to warm

sea level and sinking of the land surface are forecast to result in a 1.8-metre rise in Bangladesh in just 50 years or so. This will see the loss of a huge 16 per cent of the land surface, which supports 13 per cent of the population.

Coastal flooding will also be enhanced by storm surges, with the numbers affected predicted to rise by up to 200 million people by 2080. Because the oceans are so slow to respond to change, the problem of sea-level rise is not going to go away for a very long time. Even if we stabilized greenhouse gases in the atmosphere at current concentrations, sea level would continue to rise for a thousand years or more.

It has become fashionable to blame every weather-related natural disaster on global warming. While it is not possible to say that a specific storm or flood is due to warming, there is accumulating evidence for ever greater numbers of extreme weather events. Extreme precipitation events have increased by up to 4 per cent at high and mid-latitudes during the second half of the twentieth century, and more rainstorms, floods, and windstorms are forecast. Current climatic characteristics are likely to be enhanced, so regions that are already wet will get wetter and those that are dry will suffer from prolonged and sustained drought. Northern Europe and the UK will therefore face more floods, while the North African deserts begin to creep towards southern Europe, and Australia begins to bake beneath a blazing sun. The Atlantic's 'hurricane alley' is likely to get much busier in the next half-century, and predictions made in summer 2001 in the journal *Science* point to more strong hurricanes battering the Caribbean islands and the south-eastern and Gulf coasts of the USA. So far few are prepared to stick their necks out and say that this is definitely the result of global warming. However, as a rise in sea surface temperatures has been proposed as the driving mechanism for these more powerful storms, it would seem to be a reasonable link to make; global warming means warmer seas, which in turn are likely to give us more and bigger storms. As the tropical Atlantic has warmed over the past five years so the rate of hurricane formation has doubled. At the same time, the storms are getting stronger, with a 250 per cent increase in storms with sustained wind speeds exceeding 175 kilometres an hour. With increased warming of the oceans expected to continue throughout the twenty-first century, prospects for the inhabitants of hurricane alley look far from rosy. Where wind leads, so waves often follow, and evidence is now coming to light of bigger and more powerful waves. Around the western and southern coasts of the UK, average wave heights—about 3 metres—have risen by over a metre compared to three decades ago, while the height of the largest waves has increased by an alarming 3 metres, to 10 metres. Although not yet attributed directly to global warming, the increased wave heights reflect changes in the weather patterns of the North Atlantic that in turn can be linked to the reorganization of our planet's weather system as it continues to warm. More coastal erosion is already taking its toll along many sections of the UK's most exposed coastlines; a situation which is likely to get much worse and which will undoubtedly be exacerbated by rising sea levels and storms.

It also looks as if global warming is leading to more

12 Hurricane Andrew caused severe destruction in south Miami in 1992. More and bigger hurricanes could be on the way

frequent El Niño events; the second largest climatic 'signal' after the seasons. An El Niño involves a weakening of the westward-blowing trade winds and the resulting migration of warm surface waters from the west to the eastern Pacific, devastating local fisheries and seriously disrupting the world's climate. The World Meteorological Organization has just announced that another El Niño will be with us in 2002, bringing heavy rains to the western USA and central and southern America and drought to parts of Africa and Asia. The frequency of this particularly insidious phenomenon has risen from once every six years during the seventeenth century to once every 2.2 years since the 1970s and global warming is being held up as the culprit.

As the Earth continues to heat up, it looks as if it won't only be the seas and the skies that become increasingly agitated: the planet's crust will also join in. Already warmer temperatures in mountainous regions such as the Alps and the Pyrenees are causing the permafrost to melt at higher altitudes, threatening villages, towns, and ski resorts with more frequent and more destructive landslides. As the melting ice weakens the mountains, Switzerland is already experiencing more rockfalls, landslides, and mudflows, but things could get much worse. Whole mountainsides, consisting of billions of tonnes of rock, could collapse, burying entire communities under massive piles of rubble. Over the last 100–150 years the tops of mountains in western Europe have warmed by one or two degrees Celsius and this may be accelerating. In the mountains above the Swiss ski resort of St Moritz, for example, the temperature has risen by half a degree Celsius in just the last 15 years. Continued warming at this rate could destabilize mountain tops right across the planet, making life both difficult and dangerous for the inhabitants of high mountain terrain. A colleague of mine, Dr Simon Day, has even proposed that increasing rainfall on ocean island volcanoes may trigger gigantic landslides capable of sending huge tsunamis across the Pacific or Atlantic Oceans, but more about this in Chapter 4.

Clearly then, a major consequence of global warming will be a far more hazardous world, few of whose inhabitants will escape scot-free. Already, things are getting rapidly worse, particularly along low-lying coasts and islands. In the 1990s over 40 per cent of Solomon Islanders were either killed or impinged upon by storm and flood. Other low-lying southwest Pacific island states such as Tonga and Micronesia are also faring badly. Over the same period 1 in 12 people in Australia and 1 in 200 in the USA were hit by natural disasters, and in the UK 1 in 2,000. But this is just the start. In the first year of the new millennium, over 200 million people were affected by natural disasters—mostly flood, storm, and drought—an amazing 1 in 30 of the planet's population, and global warming has not really got going yet. Without doubt, all of us will be forced to embrace natural hazards as a normal, if unwelcome, part of our lives in the decades to come. Furthermore, the consequences of global warming stretch far beyond making the Earth more prone to natural catastrophes. Other dramatic and widespread changes are on the way that will have an equally drastic impact on all our lives. National economies will be knocked sideways and the fabric of our global society will begin to come apart at the seams, as agriculture, water supplies, wildlife, and human health become increasingly embattled.

A few countries will be able to adapt to some extent but the speed of change is certain to be so rapid as to make this all but impossible for the most vulnerable countries in Asia, Africa, and elsewhere in the developing world. Against a background of soaring populations, falling incomes, and increasing pollution, there is no question that the impact of global warming will be terrible. One of the greatest problems will be a desperate shortage of water. Even today, 1.7 billion people —a third of the world's population—live in countries where supplies of potable water are inadequate, and this figure will top 5 billion in just 25 years, triggering water conflicts across much of Asia and Africa. Alongside this, crop yields are forecast to fall in tropical, subtropical, and many mid-latitude regions, leading to the expansion of deserts, food shortages, and famine. The struggle for food and water will lead to economic migration on a gigantic scale, dwarfing anything seen today, bringing instability and conflict to many parts of the world.

In Europe and Asia trees come into leaf in spring a week earlier than just 20 years ago and autumn arrives 10 days later than it did. While this may seem beneficial, it will also encourage new pests to move into temperate zones from which they have previously been absent. Termites have already established a base in the southern UK where, in places, temperatures are now high enough for malarial mosquitoes to survive and breed. In the tropics there will be an enormous increase in the number of people at risk from insect-borne diseases, especially malaria and dengue fever, while the paucity of drinkable water will ensure that cholera continues to make huge inroads into the numbers of young, old, and infirm. In urban areas, a combination of roasting summers and increased pollution will also begin to take their toll on health, particularly—once again—in poor communities where air-conditioning is out of the question. With land temperatures across all continents due to rise by up to 8 degrees Celsius by the end of the century, temperate and tropical forests, which currently help to absorb greenhouse gases, will start to die back, taking with them thousands of animal species unable to adapt to the new conditions. And not just the forests: grasslands, wetlands, coral reefs and atolls, mangrove swamps, and sensitive polar and alpine ecosystems will all struggle to survive and adapt, and many will fail to do so. Even our leisure activities will be affected. Not only will southern Europe become too dry for cereal crops, but it will also be too hot—in the summer months at least—for sun seekers. Prospects for the winter sports industry also look bleak, with most mountain glaciers likely to have vanished by the end of the century, and snowfall much reduced. From a biodiversity point of view—as well as a tourist industry one—probably the worst recent forecast is that all the great reefs will be dead and gone within 50 years; some of the greatest natural wonders of the world obliterated by warmer seas just so that some of us can continue to live, or strive for, lives of conspicuous consumption.

Everything I have talked about so far is either already

happening or has been predicted by powerful computer-driven climate models that are constantly being upgraded in attempts to forecast better what global warming holds in store for us. We must always be prepared, however, to expect the unexpected; drastic consequences that so far have been regarded as possible but not likely, or others that have simply not been thought of. In the next chapter I will address one of these issues in more detail: the possibility of an island of cold in northwest Europe set amidst an overheated world. Here, though, I want to raise another frightening possibility—that large sea-level changes due to global warming might trigger more volcanic eruptions, earthquakes, and giant landslides. Sounds crazy? Evidence from the past suggests that it might well be possible. When sea levels were rising rapidly following the end of the last Ice Age 10,000 years ago, the weight of the water on continental margins appears to have had a dramatic effect, causing volcanoes to erupt, active faults to move, and huge landslides to collapse from continental shelf regions. The average rate of sea-level rise during post-glacial times was—at around 7 millimetres a year—just about comparable with the rise we would see should the Greenland Ice Sheet eventually succumb to global warming.

The problem is that we don't know how big or how fast a rise is needed to see these effects happening again, although, interestingly, the Pavlov volcano in Alaska is induced to erupt in winter when low-pressure weather systems passing over raise sea level by just a few tens of centimetres. Perhaps then, we face not just a warm but a fiery future. There are other worries too. The accumulation of gases from the decomposition of organic detritus leads to the formation of what are called *gas hydrates* in marine sediments. These are methane solids that look rather like water ice, whose physical state is very sensitive to changes in temperature. A warming of just one degree Celsius may cause rapid dissociation of the solid into a gaseous state, exerting increased pressure on the enclosing sediments and potentially leading to the destabilization and collapse of a huge sediment mass. This mechanism has been put forward for triggering the *Storegga Slides*—a series of gigantic submarine landslides off the coast of southern Norway—as the Earth continued to warm up 7,000 years ago. The collapses sent huge tsunamis pouring across northeast Scotland, leaving sandy deposits within the thick layers of boggy peat. If global warming really gets going and continues unhindered for the next few centuries then it looks as if things may start to get very exciting indeed.

The good, the bad, and the downright mad

No one on the planet is going to escape the effects of global warming, and for billions the resulting environmental deterioration is going to make life considerably more difficult. It is too late now to put the clock back, but we can at least attempt to alleviate the worst impacts of warming. The question is, will we ever be able to achieve a worthwhile international consensus that allows us to do this with any degree of effectiveness? The Kyoto Protocol gave us some hope in 1997, with its goal of a 5.2 per cent reduction of greenhouse gas emissions (below 1990 levels) by 2008–12, but following the failure of the USA to ratify the agreement we are back to square one. In fact, we are even worse off than this. Without US ratification, emissions from all the industrial countries put together could rise by about 12 per cent by 2008–12, which is even higher than the 'business as usual' predictions. In terms of greenhouse gas emissions, things are getting steadily worse not better. It is difficult to see how this situation can improve until the United States—the world's greatest polluter, emitting a quarter of all greenhouse gases—together with its almost equally profligate partners in crime, Australia and Canada, can be persuaded to join the rest of the international community in trying to tackle the problem. Personally, I suspect that the only persuasion that will stand any chance of working will be the persistent pounding of eastern US cities by hurricanes or perhaps a decade-long drought in Australia.

The more global warming continues to grab the limelight, so the more we hear from what I will call the 'techno-fix tendency'. Some of their proposals for mitigating warming are wild and wacky, such as placing giant reflectors in space to divert solar radiation or, even more fantastically—and heaven forbid—diverting a comet or two past the Earth, using their gravity to swing the planet out into an orbit further from the Sun. Others are seriously thought-out scientific options that we may well have to utilize at some point in the future if the situation gets really out of hand. The latter include ways of using the oceans as a dumping ground for atmospheric carbon dioxide, either by physically discarding it in the deep ocean via pipeline and tanker, or by seeding the ocean with iron to encourage the growth of marine micro-organisms that extract carbon dioxide from the atmosphere. Pilot experiments have shown that both methods can work, but to operate on a large enough scale to make any difference they would be hugely expensive and require a concerted international effort that is difficult to foresee unless the current position becomes untenable. Furthermore, convincing public opinion that we need to mess about with the oceans in order to repair the damage we have wrought in the atmosphere would be a considerable PR coup.

There is no doubt that if we are to have any impact on global warming we will all have to change our lifestyles, moving away from a disposable society and towards one that promotes and rewards the most effective and efficient use of available energy and resources. Tackling global warming is inextricably linked with the widespread adoption of sustainable development. Global warming will bring to an end the world as we have known it through dramatic changes to our environment, but if the situation is not to continue to slide it must also provide the incentive and impetus for changing the way we live. In the developed world we have no choice but to cut fuel consumption, invest in renewable energy sources, recycle on an immensely greater scale, and produce locally as much as possible rather than flying fruit and vegetables halfway around the planet. Much as I can understand their resistance, governments of developing countries must not follow the wasteful route to industrialization that Europe and North America have taken, for the simple and logical reason

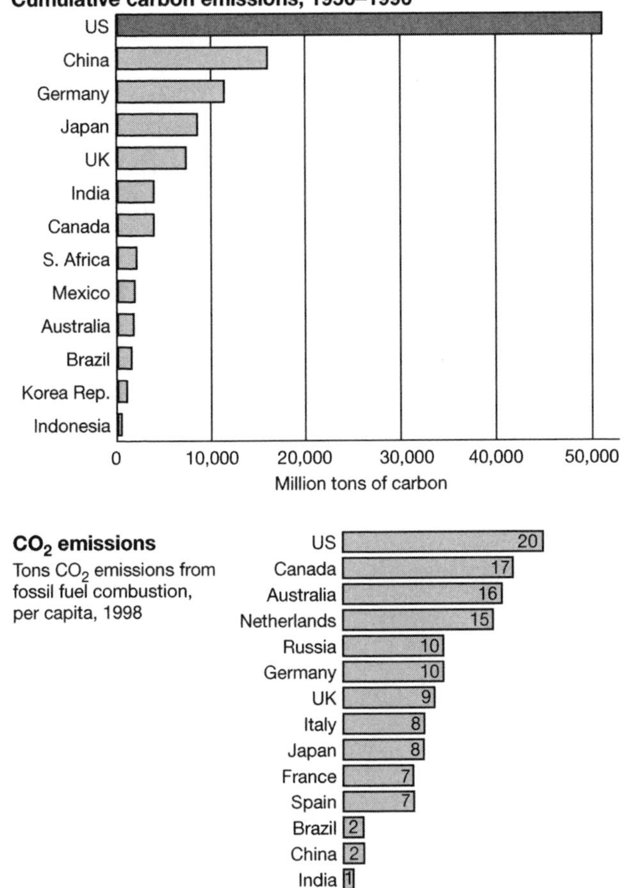

13 This country-by-country breakdown of carbon emissions reveals who the greatest polluters are

that if they don't, they—and their people—will be the ones who suffer most. In particular, the developing world has to embrace renewable energy sources and recycling now, and the world's economic powers have a duty to support them on this path. Despite the gloom after the collapse of the Kyoto Protocol, there is an alternative plan to reduce greenhouse gas emissions on the table that might just start things moving on the long road to stabilization and even reduction. Called *Contraction & Convergence,* or simply C&C, the new way forward was thought up by London's Global Commons Institute. This ingenious plan is based upon two principles. First, that greenhouse gas emissions must be reduced and, second, that the means by which this is accomplished must be fair to all. C&C therefore proposes reducing emissions on a per capita basis. International agreement will determine by how much emissions must contract each year, and then permits to emit will be allocated to all countries on the basis of their populations. The emission permits would be tradable so that countries such as the USA and Australia that could not manage within their allocations could buy extra ones from populous developing countries with a surplus. This remarkably simple scheme has not yet entered the limelight, but it does have many powerful supporters in the UN, Europe, and China, and even amongst developing countries and US senators. It is now inevitable that we and our descendants are going to face a long and hard struggle as our temperate world draws to a close and we enter the time of hothouse Earth. Perhaps, however, C&C can help to make the transition a little less desperate.

Facts to fret over

- By the end of this century the Earth is predicted to be hotter than at any time in the past 150,000 years.
- By 2100, global temperatures are forecast to rise by up to 8 degrees Celsius over land, with sea levels up to 88 centimetres higher.
- Carbon dioxide concentrations in the atmosphere may be higher than at any time in the last 20 million years.
- In the year 2000, 1 in 30 of the world's population were affected by natural disasters.
- By 2025, 5 billion people will live in countries with inadequate water supplies.
- Within 50 years all the world's great reefs will have been wiped out by higher sea temperatures.
- The winter sports industry is unlikely to survive to 2100 in its current form.
- If the Greenland Ice Sheet melts, all the world's coastal cities will be drowned, from New York to London to Sydney.

3

The Ice Age Cometh

Fire or ice?

One of the main reasons for a growing disillusionment with science amongst the general public is the perception that scientists are always arguing with one another and constantly changing their minds. It is no use explaining that this is how science progresses, through battles between competing theses until the accumulation of evidence ensures that one triumphs and becomes an accepted paradigm. People want scientists to agree, to present a united front, and to tell them what is true and what is not. They want this because it makes life that much easier and gives them that much less to worry about. If you are concerned about your career or your marriage you don't want to think about whether GM crops are good or bad, or whether you have to eat your beef on the bone or off, or whether your children's children are going to fry or freeze. Here once again, however, the scientific consensus at least appears to have done another U-turn over the last couple of decades. As we saw in the last chapter, all but the most maverick of climatologists now accept that the Earth is warming up rapidly and that our polluting activities are the cause. As recently as the 1980s, however, the big question in climatological circles was when can we expect the next Ice Age? So what has changed? Well, actually, not much. As I will explain shortly, the glaciers are still due to advance once again and we should expect our planet to be plunged into bitter cold within the next few thousand years. What has changed, however, is the recognition that anthropogenic warming and its associated climatic impact may have a role to play at a critical time of natural transition when our interglacial world is due to give itself over to ice and snow for tens of thousands of years. Problematically, however, researchers are not quite sure what this role will be, and although, intuitively, you might expect global warming to delay or even fend off entirely the next Ice Age, some scientists have suggested that the ongoing dramatic rise in temperatures may actually accelerate the onset of the next big freeze. Even if the latter is shown not to be the case, we still have a problem. Knowing that a new age of ice is on the way should we not be trying actively to keep our planet warm? Should we not welcome global warming with open arms? In other words, we are currently faced with a stark choice that is only rarely voiced during the great global warming debate. How do we wish our familiar, contemporary world to end—by fire or by ice?

How to freeze a planet

During the Earth's early history the surface boiled with lava oceans and exploding volcanoes, and although temperatures fell dramatically as prevailing geological processes moderated, our planet has been bathed in warmth for most of its 4.6 billion year history. Occasionally, however, a fortuitous combination of circumstances has heralded the formation of enormous ice sheets that have transformed a balmy paradise into a freezing hell. Artists' impressions and television documentaries have ensured that most of us are familiar with the last great Ice Age, when mammoths roamed the tundra and our pelt-covered ancestors struggled to eke out an existence from a frozen world. Only recently, however, have studies of ice-related rock formations around the world brought to light a far more ancient and much more terrible period of refrigeration; a time when our planet was little more than a frozen snowball hurtling through space. Long, long ago, during a geological episode that is becoming increasingly and appropriately referred to as *the Cryogenian* (after *cryogen* for freezing mixture), the Earth found itself at a critical threshold in its history. It had cooled substantially since its formation over three and a half billion years earlier and now the problem was keeping itself warm. At this time, between about 800 and 600 million years ago, the Sun was weaker and the Earth was bathed in some 6 per cent less solar radiation than it is now. Furthermore, the concentrations of greenhouse gases that are now heating up our planet—primarily carbon dioxide and methane—were not sufficiently high to hold back the bitter cold of space. Huge ice sheets rapidly formed and pushed towards the equator from both poles, encasing the Earth in a carapace of ice a kilometre thick. As the blinding white shell reflected solar radiation back into space, temperatures fell to -50 degrees Celsius and prospects for an eternity of ice seemed strong. But something must have happened to break the ice, as it were, otherwise I would not be here today to tell you about it; and in fact it seems that these 'snowball' conditions may have developed up to six times, succumbing each time to a return of warmer climes.

Just how the Earth managed to escape the clutches of the ice no one is quite certain, but it looks as if volcanoes might have been the saviours. After millions or even tens of millions of years of bitter cold, the enormous volumes of carbon dioxide pumped out by erupting volcanoes seem to have generated a sufficiently large greenhouse effect to warm the atmosphere and melt the ice. Extraordinarily, life came through this particularly traumatic period of Earth history bruised and battered but raring to go, and hard on the heels of Snowball Earth's final fling came the great explosion of biodiversity that marked the start of the Cambrian period 565 million years ago. Compared to the great freezes of the Cryogenian our most recent *Quaternary* ice ages come across as rather small beer. Nevertheless, although they affected smaller areas of the Earth's surface, these latest bouts of cold were crucial because they coincided with the appearance and evolution of our distant ancestors. Furthermore, they may yet have a role to play in the future of our race.

During recent Earth history the Sun's output has been significantly higher than during the Cryogenian and the level of carbon dioxide and other greenhouse gases has also been higher. Why then, at the end of the Miocene period about 10 million years ago, did glaciers once again begin to form and advance across parts of the northern hemisphere? And more importantly, why, around 3 million years ago, did the southward march of the ice intensify? This remains a particularly

14 Snowball Earth: artist's impression of an ice-covered planet. During the Earth's early history, ice advanced from the poles to transform the planet into a great snowball

hot topic in the fields of Quaternary science and environmental change and a detailed analysis of competing theories is beyond the scope of this book. Suffice it to say that explanations for the twenty or so ice ages that have gripped the Earth during the last 2 million years include disruption of the planet's atmospheric circulation due to uplift of the great Himalayan mountain belt, and the drastic modification of the global system of ocean currents by the emergence of the Panama Isthmus.

Although one or both of these spectacular geophysical events may have contributed to a picture of increasing cold, the ice was already on the move, and we need to look elsewhere for the true underlying cause. What, in other words, turns ice ages on, and—just as importantly—what turns them off? This problem has intrigued scientists for many years and the solution was first put forward by the Scottish geologist James Croll as long ago as 1864 and expanded upon by the Serbian scientist Milutin Milankovitch in the 1930s. The Croll-Milankovitch *astronomical theory of the ice ages* proposes that long-term variations in the geometry of the Earth's orbit and rotation are the fundamental causes of the blooming and dying of the Quaternary ice ages. In order for an ice age to get going, the astronomical theory requires that summers at high latitudes in the northern hemisphere are sufficiently cool to allow the preservation of winter snows. As more and more snow and ice accumulates year on year, so the reflectivity or *albedo* of the surface is increased, causing summer sunshine to have even less impact and accelerating the growth of ice sheets and glaciers. But how are the northern hemisphere summers cooled down in the first place? This is where the astronomy comes in. Cooler summers at high latitudes result from a reduction in the amount of solar radiation falling on the surface, and this in turn depends upon both changes in the tilt of the Earth's axis *and* variations in its orbit about the Sun.

If the Earth's axis was not tilted then we would not experience the seasons. During the northern hemisphere summer, for example, the North Pole is tilted towards the Sun, allowing more direct solar radiation to reach the surface in the northern hemisphere and raising the temperatures. In contrast, during the winter, the North Pole is tilted away from the Sun and the long, balmy days of summer are replaced by the cold and dark of a northern hemisphere winter. Now the southern hemisphere receives more direct sunlight with the result that those down under bask in warmth while the north shivers beneath gloomy skies. Although the tilt of the Earth's axis averages about 23.5 degrees, it is not constant. Like a spinning top, the Earth wobbles—or *precesses*—about its axis of rotation over a period of between 23,000 and 26,000 years. Furthermore, this wobble causes the amount of tilt to vary between 22 and 25 degrees over a period of 41,000 years. At times of least tilt, winters are actually milder, but more importantly, high latitudes receive less direct solar radiation and become cooler, making the survival of winter snows and the growth of ice sheets easier. On top of this there is another so-called astronomical *forcing* mechanism that contributes to the onset of ice age conditions. Like all planetary bodies, the Earth follows an elliptical rather than a circular path around the Sun, whose shape varies according to cycles of between 100,000 and 400,000 years. At the moment the Earth's closest approach to the Sun occurs in January, when the North Pole is pointing away from the Sun, resulting in slightly colder northern hemisphere winters. Just 11,000 years ago, however, this closest approach—or *perihelion*—occurred in July, giving a small temperature boost to northern hemisphere summers.

Before this gets too complicated let me try and draw things together. Regular and predictable cycles—known as *Milankovitch Cycles*—are recognized in the behaviour of the Earth's tilt and its orbit over periods of thousands to hundreds of thousands of years, and these cycles control the amount of solar radiation reaching the Earth's surface and therefore its temperature. At times, a number of cycles may coincide so as to depress summer temperatures at high latitudes to a degree sufficient to allow the accumulation of winter snows. On its own this could not result in the huge ice sheets that have dominated the northern hemisphere for much of the last few million years, but as the area covered by snow and ice grows, so more and more sunlight is reflected back into space, accelerating the cooling process. This—in essence—is how ice ages start. Conversely, at other times, the various cycles cancel one another out, the planet warms as a result, and the ice sheets retreat to their polar fastnesses.

Although Milankovitch and later researchers who have addressed the issue have been able to explain the mechanics of the ice ages and their periodicity, they have been less successful in deciding why these icy episodes appeared on

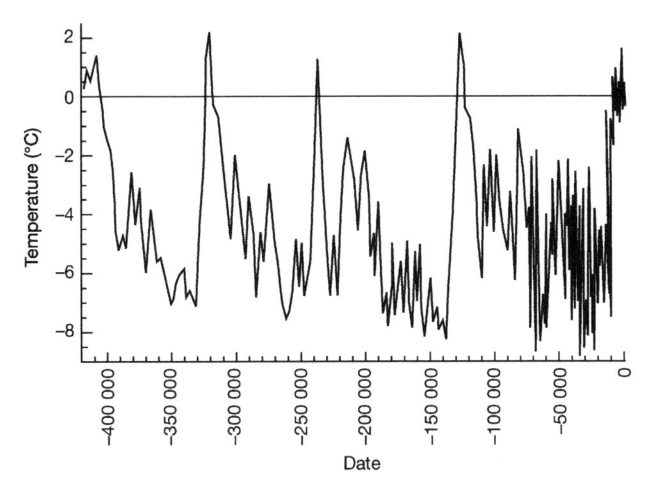

15 Temperature changes over the past 420,000 years show that the Earth has been much colder than it is now for most of the time

the scene around ten million years ago, rather than being apparent throughout Earth history. An answer to this may lie, however, in the carbon dioxide level of the Earth's atmosphere, which has been steadily falling over the last 300 million years, from about 1,600 parts per million to just 300 ppm prior to the industrial revolution. It has been suggested that perhaps only when the level of carbon dioxide in the atmosphere drops below a critical threshold level—say of 400 ppm—is astronomical forcing sufficient to initiate the cycle of warm and cold that characterizes the ice ages. This begs the question that with carbon dioxide levels expected to rise above this level in a little over 20 years, have we seen off the ice ages for ever? I shall return to this later.

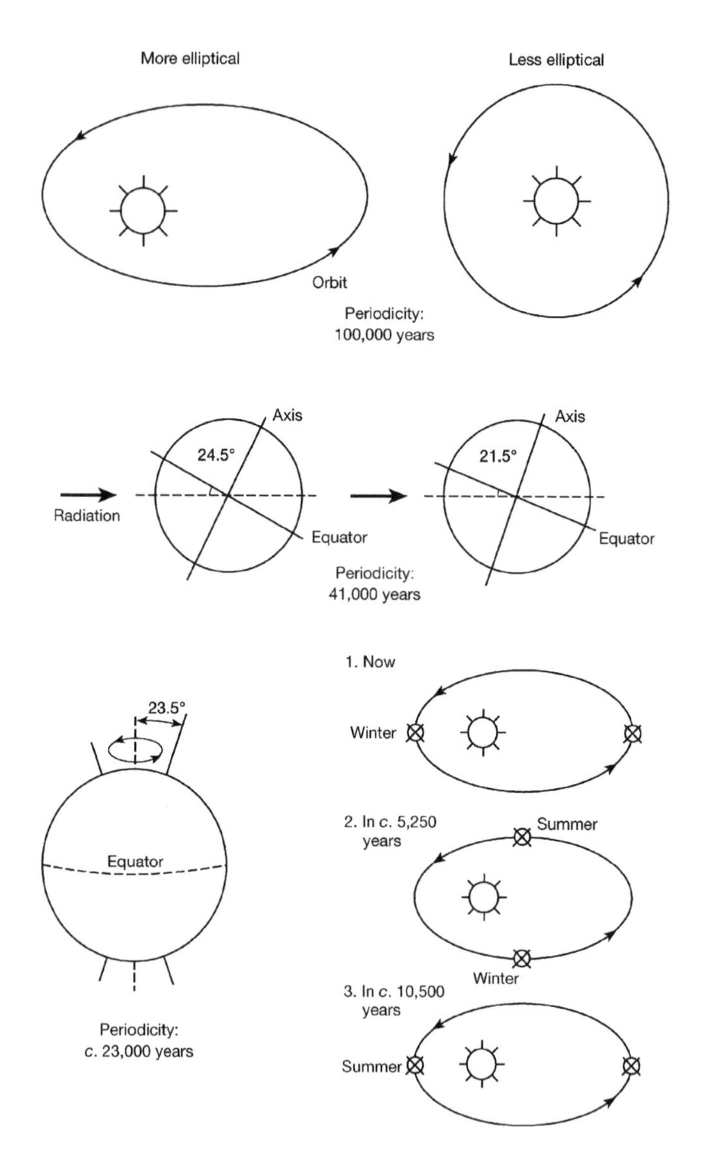

16 Milankovich cycles control the timing of the ice ages; variations in the shape of the Earth's orbit around the Sun (top), changes in the tilt of the Earth's axis (middle), and precession about the Earth's axis (bottom)

In the meantime, on the basis that there is at least a fair chance that we will have to face them again at some time in the future, let's examine what conditions were like in the depths of the last Ice Age. As temperatures started to fall around 120,000 years ago, so more and more of the planet's water found itself locked up in mountain glaciers, polar sea ice, and expanding continental ice sheets in the northern hemisphere, with the result that sea level began to fall dramatically. Ice swept southwards towards the equator on at least four occasions over this period, with the peak of ice cover being reached a mere 15,000–20,000 years ago.

At this time sea level was a good 120 metres or so below what it is now—about the height of a 40-storey building—exposing new land bridges between continents and permitting the migration of both animals and our distant ancestors. One of these land bridges developed across the Bering Straits, allowing people from Asia to cross into North America, from which, eventually, they colonized the New World. Just 600 generations ago, then, the north of our planet was in the steely grip of full glaciation with a third of all land covered by ice and 5 per cent of the world's oceans frozen. Compared to today, the environment at the height of the last Ice Age was desperately hostile, with mean temperatures 4 degrees Celsius lower than today but far lower at high latitudes in the north. In the UK, temperatures were reduced by between 15 and 20 degrees Celsius, transforming the country into a frozen wasteland with great sheets of ice reaching as far south as the River Thames and beyond. Some of the most inhospitable conditions were, however, to be found in North America, where temperatures over huge areas were 25 degrees Celsius lower than today and ice fields kilometres thick ensured that life was largely impossible. Remarkably, however, just as it seemed as if the world might be returning to the 'snowball' state of the Cryogenian, a surprising change took place. The planet started to warm rapidly, melting the great ice sheets at a rate far quicker than it took them to form. Meltwater poured into gigantic lakes at the margins of the ice fields, which, in turn, emptied into the oceans, raising sea level and inundating land exposed just a few thousand years earlier. By 12,000 years ago, sea level was rising far more rapidly than even the most pessimistic forecasts for the next century, possibly by as much as 10 metres or so in a couple of centuries, and all the time the climate was becoming warmer and warmer—well, almost all the time that is. The journey from the depths of ice age to the current balmy interglacial was a rather bumpy one, and on more than one occasion the ice made a concerted attempt to reclaim centre stage. Around 11,000 years ago, for example, the rapid retreat of the ice was stopped in its tracks as a new blast of cold initiated a thousand-year-long freeze, known as the Younger Dryas to distinguish it from an earlier and less severe cold phase called the Older Dryas. No one is certain what caused this sudden cold snap but one suggestion is that the culprit was a huge discharge of fresh water from long-gone Lake Agassiz, one of the gigantic glacial meltwater lakes that had accumulated in North America. The catastrophic emptying of this lake into the St Lawrence, and thence into the

17 Ice covered huge areas of the Northern Hemisphere land masses during the last Ice Age

North Atlantic, may have disrupted currents carrying warmer waters into polar regions, allowing the climate at higher latitudes to cool and ice to form once again. The Younger Dryas and similar post-Ice Age cold snaps teach us a number of important lessons that we would do well to remember as our own world undergoes dramatic climate change. First, the switch from warm to cold and vice versa can occur extraordinarily rapidly—within decades—and second, the disruption of ocean currents can have serious and far-reaching consequences for climate change. Some worrying implications of the latter I will address in more detail later in this chapter.

Charles Dickens, White Christmases, and the Little Ice Age

It seems likely that almost everyone who has read this far is familiar with *the* Ice Age, but what about the *Little* Ice Age? This is the term used by climatologists to describe a cold period that lasted from at least 1450—and possibly 1200—until between 1850 and the start of the twentieth century. Over this period, glaciers advanced rapidly, engulfing alpine villages, and sea ice in the North Atlantic severely disrupted the fishing industries of Iceland and Scandinavia. Eskimos are alleged to have paddled as far south as Scotland, while the once thriving Viking community in Greenland was cut off and never heard from again.

Annual mean temperatures in England during the late seventeenth century were almost one degree Celsius lower than for the period 1920–60, leading to bitter, icy winters in which 'frost fair' carnivals were held regularly on a frozen River Thames and snowfall was common. The snowy winters described in many of the works of Charles Dickens may well be a reflection of this colder climatic phase, and they have certainly done much to nurture our constant expectation of—and wish for—an old-fashioned 'white Christmas'.

Just what was the cause of the Little Ice Age remains a matter of intense debate. Clearly, however, as most of the cold snap occurred prior to the industrial revolution there can be no question of human activities having played a role. Despite this, it is vital that we understand the Little Ice Age in the context of global warming, if for no other reason than if we don't appreciate the natural variations of our planet's recent climate it is well nigh impossible to unravel the effects arising from human activities. In fact, the Little Ice Age was not the only significant departure from the climatic norm—if there is such a thing—in historical times. Immediately prior to this cold snap, Europe, at least, was revelling in the so-called *Medieval Warm Period.* This time of climate amelioration, between about 1000 and 1300 AD, saw grapes grown in England 300 miles further north than at present, while the Norse settlers of Greenland were able to graze their livestock in areas that are now buried beneath ice. The emergence of the world from the Little Ice Age towards the end of the last century, coincident with the acceleration of industrialization on a global scale, has contributed in no small way to current arguments on the causes of contemporary warming.

As I noted in the last chapter, the overwhelming scientific consensus views global warming as being anthropogenic in nature, but some still hold out for an entirely natural cause; seeing the current warming in terms of the planet coming out of the Little Ice Age and entering another warm period analogous to the Medieval Warm Period. Although the available evidence irresistibly supports a human cause rather than a purely natural warming, there can be no question that the impact of human activities is superimposed upon a natural variability that in the recent past has resulted in significant climate change. But what is the cause? One of the most likely culprits is the Sun, whose output continues to vary on time scales ranging from 100 to 10,000 years. For example, the two coldest phases within the Little Ice Age corresponded closely with two periods of apparently reduced solar activity; the Spörer Minimum between 1400 and 1510 AD and the Maunder Minimum from 1645 to 1715. During these periods, virtually no sunspots were visible and auroras were almost non-existent, suggesting a fall in the rate of bombardment of the Earth by solar radiation. While solar physicists estimate that the Sun during the Maunder Minimum may have been just a quarter of one per cent dimmer than it is today, this might have been sufficient to cause the observed cooling. Other factors may also have made a contribution, however, and a recent theory has given elevated levels of explosive volcanic activity at the time—including the great 1815 eruption of Indonesia's Tambora volcano—at least a supporting role in the Little Ice Age cooling. As I will discuss further in the next chapter, large volcanic explosions are particularly effective at injecting substantial volumes of sulphur dioxide and other sulphur gases into the stratosphere—that

18 During the Little Ice Age winters were often cold enough for ice fairs on the River Thames

part of the atmosphere above 10 kilometres or so. Here they mix with atmospheric water vapour to form a fine mist of sulphuric acid that cuts out a proportion of incoming solar radiation and leads to a cooling of the troposphere (the lower atmosphere) and surface.

A very British ice age

The more we learn about past climate change, the more it becomes apparent that dramatic variations can occur with extraordinary rapidity. The return—possibly within a few decades—from increasingly clement conditions to the bitter cold of the Younger Dryas, 11,000 years ago, demonstrates this, as does the similarly rapid transition from the Medieval Warm Period to the Little Ice Age. Equally disturbing is the tendency for the climate to flip suddenly from one extreme to another when it is under particular stress, as it is at the moment from anthropogenic warming. This once again raises the question I posed at the beginning of this chapter—is there any way that current global warming can actually bring a return to colder conditions? While this would seem to be counter-intuitive, there is increasing evidence that this may well happen—at least as far as the UK and northwest Europe are concerned. The only reason why it is possible for tropical palms to thrive in western Ireland and southwest England is because the Gulf Stream carries northwards warm water from the Caribbean. As a result, the UK and Ireland are substantially warmer than comparable latitudes in eastern Canada, which have to put up with sub-Arctic conditions. But what would happen if the supply of warm water from the south were shut down? It is highly likely that the British climate—and perhaps that of much of northwest Europe—would become bitterly cold, and some have suggested it could even rival that of Svalbard (formerly Spitsbergen), the ice-shrouded islands off east Greenland where the polar bear is king.

One of the ways of weakening or shutting down the Gulf Stream is by short-circuiting it through releasing huge quantities of cold fresh water into the North Atlantic, and this is just what is predicted by a number of different climate models developed to look at the impact of global warming in this century and beyond. If greenhouse gas emissions double over the next 70 years then the warming currents could well weaken by up to 30 per cent. Even worse, under a 'business as usual' scenario, greenhouse gas emissions will quadruple by the end of the century, leading—according to one model—to a complete shutdown of the supply of warm water to the northeast Atlantic by the middle of the next century.

In little more than half a century, then, the seas around the UK could be significantly cooler, altering prevailing weather patterns and bringing colder conditions to the region. While the rest of the world roasts, the UK and northwest Europe could conceivably start to slide into a freeze very much more bitter than the Little Ice Age. And this might be just the start. The knock-on effects of changes to the ocean circulation in the North Atlantic may spread, overwhelming the current warming and bringing a return of the ice across the northern hemisphere. In conclusion, then, let's take a look at prospects for the return of the Ice Age and the role mankind may already be playing in its reappearance.

Out of the frying pan into the fridge

In terms of the Milankovitch Cycles, our planet is already primed for the end of the current interglacial and a return to full Ice Age conditions. Some believe that all that is needed is a trigger; a sudden shock to the system that will knock the climate out of equilibrium and set it wobbling before it collapses into an altogether less friendly state. It is questionable whether global warming can provide a shock of the appropriate magnitude, but new research is leading to increasing concern that the legacy of warming today may be freezing tomorrow. Once again, the key seems to lie in the ocean circulation system of the North Atlantic, which appears to be closely bound up with past switches from warm to cold episodes and vice versa. The Gulf Stream that most people are familiar with is actually only one part of a system of currents known by a variety of names, of which the *Atlantic Overturning Circulation* is probably the most revealing. As the warm waters of the Gulf Stream head northwards they cool and consequently become more dense. As a result, by the time they have reached the Arctic Ocean they have sunk to form a cold, deep-ocean current that heads south once more to join the wider system of ocean currents known as the *Global Conveyor.*

It now looks as if operation of the Atlantic Overturning Circulation is seriously disrupted whenever cold conditions grip the northern hemisphere. During the Younger Dryas, for example, the circulation appears to have been severely reduced, lowering north European temperatures by as much as 10 degrees Celsius. Recent evidence on ocean temperatures and salinities, gleaned from studies of the shells of tiny marine organisms known as *foraminifera,* also points to a much weaker Gulf Stream at the height of the last Ice Age some 20,000 years ago. Then, it seems, the Gulf Stream had only two-thirds of its current strength, suggesting that the entire circulation system was comparably weakened. The question is, did this weakening have a role to play in the triggering of the last Ice Age, or was it merely a consequence?

No one really knows, but there is a general feeling that a weakening of the circulation results in much colder conditions in the northern hemisphere and that such a weakening appears to be associated with large influxes of cold water into the North Atlantic. Due to melting of Arctic sea ice and the Greenland Ice Cap, this is just what is predicted to happen in the next few centuries.

During the Younger Dryas, 11,000 years ago, the release of huge quantities of water from glacial lakes resulted only in a short-lived cold snap of a thousand years or so. Then, however, the Earth was at a point in the pattern of Milankovitch Cycles when temperatures were on the way up. Now, we are poised at the transition between the present interglacial and the next Ice Age, and without the polluting effects of human activities temperatures could be expected to be on the way down. It is not unreasonable to at least consider, then, that the influx of cold, fresh water into the Arctic Ocean may trigger not just a brief period of cold in northwest Europe, but a new Ice Age affecting the entire northern hemisphere. And we may not have too long to wait. In the 1990s, US climate modellers Ronald Stouffer and Alex Hall ran a comprehensive computer model of the Earth's climate system for almost a decade to find out what it had in store for us in the next few millennia. What they discovered was seriously disturbing. The model predicts that, in around 3,000 years' time, intense westerly winds over Greenland will help to push large quantities of fresh Arctic water into the North Atlantic. Because of its low density, this bitterly cold water will remain at the surface, cooling the air above, and creating a low-pressure weather system that will reinforce the westward gales through a positive feedback mechanism. The effect is forecast to cool the North Atlantic by up to 3 degrees Celsius and also to weaken the Atlantic Overturning Circulation, bringing colder conditions to northwest Europe. In the model, the chilly scenario only persists for 40 years or so, but the authors are concerned that if global warming promotes the melting of Greenland ice on a grand scale, this added input of cold water might amplify a brief regional cooling into a widespread and persistent freeze. Even more worryingly, the first signs of the coming chill may already have been detected, with recent measurements revealing that an important current running south between Scotland and the Faeroe Islands has slowed by around 20 per cent in the past 50 years. Could this be the first evidence of the breakdown of the Atlantic Overturning Circulation and the slow but steady deterioration of the climate into bitter cold?

One of the best means of illustrating just what a bad time this is for us to be experimenting with the global climate is by comparing the temperature profile of this interglacial with that of the last. It is rather sobering to see that the natural temperature trend is already downwards, and in fact this fall has been going on for several thousand years. At the moment, it looks as if the downward trend is being reversed by anthropogenic warming, and without greenhouse gas emissions the world would be around 3 degrees colder in around 8,000 years' time—well on its way to the next Ice Age. Although fending off the chill at the moment, however, the impact of global warming on the Atlantic Overturning

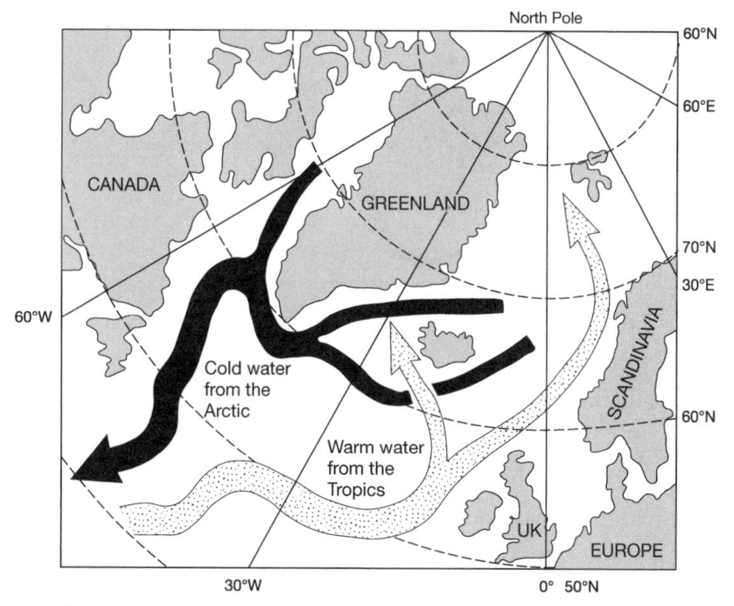

19 Warm water from the tropics heads towards the pole to be replaced by cold Arctic waters heading south

Circulation might well ultimately accelerate the arrival of the next Ice Age.

By now I hope to have convinced you that it is at least feasible for the current global warming to trigger colder conditions, and that this may be the result of the continued and unmitigated emission of greenhouse gases. So what happens if the world comes to its senses and we cut back significantly on the amount of carbon dioxide and other gases that we pump into the atmosphere? Well, you have seen

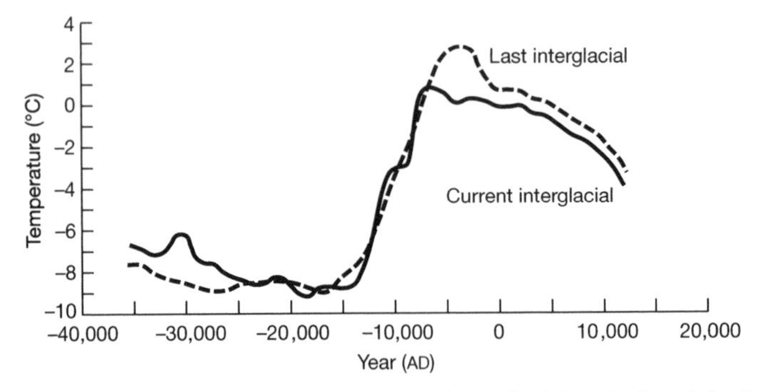

20 Comparisons of temperatures in this interglacial period and the last suggest that we are already well on the way to the next ice age

the graphs—the ice will get us anyway. It is simply just a matter of whether we want to take the icy plunge on its own or spend some time baking in the sauna first. Whichever choice we make, there is no denying that life for our descendants will become increasingly hard, should the ice return. Life in Europe, North America, Russia, and central and eastern Asia will be pretty much impossible, engendering mass migrations southwards accompanied undoubtedly by bloody wars fought over living space and resources. The climate of Ice Age Earth is simply not suited to sustaining a population totalling 8–10 billion, or thereabouts, and widespread famine alongside civil strife is certain to lead to a severe culling of the human population. There is no question that our race will survive, as it did the last time that the ice left its polar fastnesses, but it is likely to be but a pale shadow of its former self.

Facts to fret over

- Between 800 and 600 million years ago, the Earth was a frozen snowball covered with ice a kilometre or more thick.
- Just 600 human generations have passed since the end of the last Ice Age.
- At the height of last Ice Age, temperatures in the UK were 15–20 degrees Celsius lower than they are now, and over much of North America, more than 25 degrees Celsius lower.
- Sea levels have risen by over 120 metres since the ice started to retreat around 18,000 years ago.
- Unmitigated greenhouse gas emissions may lead to a shutting off of the Atlantic Overturning Circulation within a century and a half.
- An Atlantic current flowing between Scotland and the Faeroe Islands has weakened by 20 per cent in the last 50 years.
- Without greenhouse gas emissions, the world could be 3 degrees Celsius colder in 8,000 years.

4

The Enemy Within

Super-Eruptions, Giant Tsunamis, and the Coming Great Quake

Hell on Earth

Imagine the worst possible vision of hell. The vile stench of sulphurous gas pervading a world of darkness broken only by a dull red glow on a distant, invisible horizon. Heavy, grey ash pours from above like snow, clogging the eyes, nose, and ears as swiftly as you can remove it. Choking and retching you ram your fingers into your mouth to try and gouge out the ashy, gritty slime that forces its way in with every struggling breath, but to no avail. Suddenly a blinding flash reveals the nightmare landscape of Tolkien's Mordor—all familiar features blotted out and buried by ash accumulating at half a metre an hour. A titanic crash of thunder heralds the return of the darkness and the onset of a truly biblical deluge. Within seconds the ashy drifts are transformed into rushing torrents of mud that almost sweep your feet from under you. As the falling rain and ash combine, you are battered by pellets of mud that begin to weigh you down under a sticky, ever-thickening carapace of Vulcan's ordure. There is no sign that the Sun ever bathed the landscape in its warming rays, but it is far from cold. In fact, your body is slowly roasting in the stifling heat of nature's own oven, your sweat sucking you dry as it drains from every pore to mix with the muddy rivulets covering every inch of your skin.

Some of the half a billion inhabitants of the danger zones

around the world's 500 or so historically active volcanoes don't need to use their imagination. They have already experienced hell. Show the above description to survivors from the 1991 eruption of Pinatubo (Philippines) or the twin eruptions of Vulcan and Tavurvur at Rabaul (Papua New Guinea) three years later, and they will nod their heads and say 'I have been there!' However awful it might sound to those of us who live far from the slopes of an erupting volcano, there is nothing unusual about the above scene. But what if it was enacted 1,500 kilometres from the eruption? Then it really would be something very special, because it would mean that the Earth was being rent by one of nature's greatest killers—a volcanic *super-eruption.* These gigantic blasts dwarf even the greatest eruptions of recent times, and in comparison the cataclysmic detonation that blew Krakatoa (Indonesia) apart in 1883, killing around 36,000 of the inhabitants of Java and Sumatra, pales into insignificance. Even the titanic blast that tore the Greek island of Thera to pieces one and a half millennia before the birth of Christ (thereby engineering the demise of the Minoan civilization and launching the enduring legend of Atlantis) would be little more than a firecracker alongside such an Earth-shattering event.

Fortunately for us, super-eruptions are far from common, and it is estimated that throughout the last two million years of Earth history, there have been perhaps two such blasts every hundred millennia. While natural phenomena never stick rigorously to a timetable, it is nevertheless slightly disconcerting—given this frequency—that the last such

21 Ash from the erupting Tavurvur volcano continues to fall across the town of Rabaul in New Britain (Papua New Guinea)

cataclysm occurred a good 73,500 years ago. The really scary thing, however, is that, unlike 'normal' volcanic blasts, there is no possibility of avoiding the devastating consequences of a volcanic super-eruption. Those of us tucked away in the most geologically friendly countries will still find our cosy world turned upside down by the next super-eruption, even if it occurs in a distant land on the other side of the planet. This is because of the severe impact it will have on the climate, the ash and gas ejected high into the atmosphere dramatically reducing the solar radiation reaching the surface and triggering a freezing *volcanic winter* worldwide.

Before examining the truly terrifying consequences of the next volcanic winter, let me take a more detailed look at the scale of volcanic super-eruptions, compared with the common-or-garden variety of volcanic blast. A number of scales have been devised in recent years to allow the sizes of volcanic events to be compared. One of the earliest and most commonly quoted is the *Volcanic Explosivity Index* or VEI devised by volcanologists Chris Newhall and Steve Self in 1982, primarily to allow estimation and comparison of the magnitudes and intensities of historical eruptions. *Eruption magnitude* refers to the mass of material erupted, while *eruption intensity* is a measure of the rate at which material is expelled. The index is logarithmic (like the better-known Richter Scale for earthquakes) which means that each point on the scale represents an eruption ten times larger than the one immediately below. Thus a VEI 5 is ten times larger than a 4, a VEI 6 a hundred times larger, and a VEI 7 a thousand times larger. At the bottom of the index, the gentle effusions of lava that characterize most eruptions of Kilauea and Mauna Loa on Hawaii score a measly 0, while mildly explosive eruptions that release sufficient ash to perhaps cover London or New York in a light dusting would register at 1 or 2. To a volcanologist, however, things don't really start to get exciting until higher values are reached. VEI 3 and 4 eruptions are described, respectively, as 'moderate' and 'large'. This translates into blasts big enough to cause local devastation, sending columns of ash up to 20 kilometres into the atmosphere and burying the surrounding landscape under piles of volcanic debris a metre or more deep. In 1994, the town of Rabaul in New Britain (Papua New Guinea) was destroyed by an eruption of this size, and a few years later—in 1997—Plymouth, the capital of the Caribbean island of Montserrat, suffered the same fate. Eruptions that score a 5 on the scale, such as the much-televised 1980 blast of Mount St Helens (Washington State, USA) typically cause mayhem on a regional scale, while VEI 6 eruptions can be regionally devastating and the effects long-lasting. The 1991 Pinatubo eruption in the Philippines was probably the largest eruption of the twentieth century, ejecting sufficient ash and debris to bury central London to the depth of a kilometre and making hundreds of thousands homeless. For years afterwards, mudflows continued to pour down the flanks of the once-again dormant volcano, clogging rivers, burying farmland, and flooding towns and cities. For the last VEI 7 eruption we have to go back almost two centuries to 1815—the year of the battle of Waterloo. As the armies of Wellington and Napoleon jostled for position across Europe, on the distant

Indonesian island of Sumbawa, the long-dormant volcano Tambora ripped itself apart in a gargantuan eruption that may have been the largest since the end of the Ice Age 10,000 years ago. Sir Stamford Raffles, the then British Lieutenant Governor of Java, reported a series of titanic detonations loud enough to be heard in Sumatra 1,600 kilometres away. When the eruption ended, after 34 days, it left 12,000 dead. In the ensuing months, however, a further 80,000 Indonesians succumbed to famine and disease as they struggled to find food and uncontaminated water across the ash-ravaged landscape.

Utterly devastating though the Tambora event no doubt was to the people of Indonesia, its direct effects were nonetheless confined to one part of South East Asia. Indirectly, however, much of the world was to suffer the consequences of this huge blast. Along with some 50 cubic kilometres of ash, the climactic explosions of the Tambora eruption also lofted around 200 million tonnes of sulphur-rich gases into the stratosphere, within which high-altitude winds swiftly spread them across the planet. The gases combined readily with water in the atmosphere to form 150 million tonnes of sulphuric acid *aerosols*—tiny particles of liquid that are very effective at blocking out solar radiation. Within months the northern hemisphere climate began to deteriorate and temperatures fell to such a degree that 1816 became known as the 'year without a summer.' Global temperatures are estimated to have fallen by around 0.7 degrees Celsius—perhaps a seventh of the drop required to plunge the planet into full ice age—causing summer frosts, snows, and torrential rains.

The miserable weather conditions may have set just the right mood for Mary Shelley's vivid imagination to spawn its most famous offspring, *Frankenstein*, while the spectacular ash and gas-laden sunsets are said to have inspired some of J. M. W. Turner's most brilliant works.

Certainly the weather conditions in Europe and North America during 1816 were awful, but could a volcanic eruption in a far-off part of the world really change the climate so much as to cause a breakdown in society and end the world as we know it? Evidence from the past suggests that there is no doubt that it can. Far back in the geological record—during the Ordovician period some 450 million years ago—an enormous volcanic explosion in what is now North America ejected sufficient ash and pyroclastic flows that, if it happened today, it would obliterate everything over an area of at least a million square kilometres. This is broadly the size of Egypt or four times the area of the UK. In addition the amount of gas and debris pumped into the atmosphere must have been phenomenal. A little nearer our time, just 2 million years ago, a mighty eruption at Yellowstone in Wyoming was violent enough to leave behind a gigantic crater (or *caldera)* up to 80 kilometres across, and pump out ash that fell across 16 states. Another huge eruption occurred at Yellowstone around 1.2 million years ago and yet another just 650,000 years ago. If this last cataclysm occurred today it would leave the United States and its economy in tatters and the global climate in dire straits.

The eruption scoured the surrounding countryside with hurricane-force blasts of molten magma and incandescent gases—known as *pyroclastic flows*—with a volume sufficient to cover the entire USA to a depth of 8 centimetres. Ash fell as far afield as sites that are now occupied by the cities of El Paso (Texas) and Los Angeles (California), and Yellowstone ash from this eruption is even picked up in deep-sea geological cores from the Caribbean seabed. Although no eruptions have been recorded at Yellowstone for 70,000 years, the hot springs, spectacular geysers, and bubbling mud pools provide testimony that hot magma still resides not far beneath the surface. This is further supported by the numerous earthquakes that regularly shake the region and the periodic swelling and subsiding of the land surface. No one knows

22 Hot springs, bubbling mudpools, and spectacular geysers testify to magma lurking not far beneath the surface of Yellowstone Park in Wyoming, USA

when—or even if—Yellowstone will experience another devastating super-eruption. It is a little worrying, however, to note that these huge blasts seem to occur every 650,000 years or so. Perhaps then, we are due another any time now?

It would be easy to sit back and say—that's all very well, but these horrific events took place deep within the mists of time. Surely they can't happen today? Thinking along these lines would be a very big mistake. In 186 AD a massive eruption at New Zealand's Lake Taupo ejected pyroclastic flows that devastated a substantial portion of the North Island. 73,500 years ago—considerably older but still well within the time span of modern humanity—perhaps the greatest volcanic explosion ever tore a hole 100 kilometres across at Toba in northern Sumatra. This huge caldera, which is now lake filled, is very much a tourist attraction, but there is evidence of a much more sinister legacy. The eruption of Toba may have come within a hair's breadth of making the human race extinct. Estimates of the size of the blast vary, but there is no question that—along with the Yellowstone eruptions—Toba qualifies as a VEI 8 super-eruption. It was thought that the total amount of debris ejected during the eruption was of the order of 3,000 cubic kilometres, sufficient to cover virtually the whole of India with a layer of ash one metre thick. Recent evidence from deep-sea geological cores suggests, however, that the eruption might have lasted longer than previously thought and ejected considerably more debris, perhaps up to 6,000 cubic kilometres. Almost unbelievably, this would be enough to bury the entire United States to a depth of two-thirds of a metre.

Any of our ancestors living on Sumatra at the time would without question have been obliterated. For the human race as a whole to suffer the threat of extinction, though, the effects of the eruption would have to have been severe across the whole planet, and this they seem to have been. Along with the huge quantities of ash, the Toba blast may have poured out enough sulphur gases to create up to 5,000 million tonnes of sulphuric acid aerosols in the stratosphere. This would have been sufficient to cut the amount of sunlight reaching the surface by 90 per cent, leading to global darkness and bitter cold. Temperatures in tropical regions may have rapidly fallen by up to 15 degrees Celsius, wiping out the sensitive tropical vegetation, while over the planet as a whole the temperature drop is likely to have been around 5 or 6 degrees Celsius, broadly the equivalent of plunging the planet into full ice age conditions within just a few months. Temperature records from Greenland ice cores suggest that the eruption was followed by at least six years of such volcanic winter conditions, which were in turn followed by a thousand-year cold 'snap'. Soon afterwards the planet entered the last Ice Age, and there is some speculation that in this respect, the cooling effect of the Toba eruption may have been the final straw, tipping an already cooling Earth from an interglacial into a glacial phase from which it only fully emerged around 10,000 years ago.

What then of our unfortunate ancestors: could this period of volcanic darkness and cold really have brought them to their knees? It certainly seems possible. Studies of human DNA contained in the sub-cellular structures known as mito-

23 The gigantic eruption of Toba 73,500 years ago excavated a crater 100 kilometres long and plunged the world into the depths of volcanic winter

chondria reveal that we are all much too similar—genetically speaking—to have evolved continuously and without impediment for hundreds of thousands of years. The only way to explain this extraordinary similarity is to invoke the occurrence of periodic *population bottlenecks* during which time the number of human beings was, for one reason or another, slashed and the gene pool dramatically reduced in size. At the end of the bottleneck, all individuals in the rapidly expanding population carry the inherited characteristics of this limited gene pool, eventually across the entire planet. Mike Rampino, a geologist at New York University, and anthropologist Stanley Ambrose of the University of Illinois have proposed that the last human population bottleneck may have been a consequence of the Toba super-eruption. They argue that conditions after the Toba blast would have been comparable to the aftermath of an all-out nuclear war, although without the radiation. As the soot from burning cities and vegetation would result in a *nuclear winter* following atomic Armageddon, so the billions of tonnes of sulphuric acid in the stratosphere following Toba would mean per-

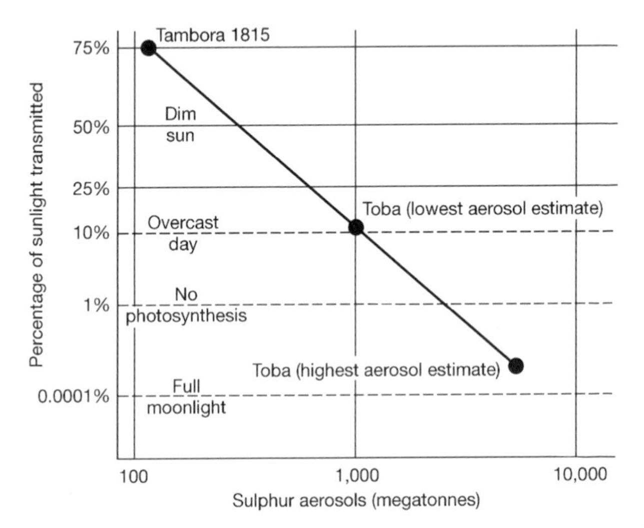

24 Sunlight reduction due to Toba: worst-case estimates suggest that sulphuric acid aerosols from Toba may have cut out so much sunlight that the entire Earth was as dark as on a night of full moon

petual darkness and cold for years. Photosynthesis would slow to almost nothing, destroying the food sources of both humans and the animals they fed upon. As the volcanic winter drew on, our ancestors simply starved to death leaving fewer and fewer of their number, perhaps in areas sheltered for geographical or climatological reasons from the worst of the catastrophe. It has been suggested that for 20 millennia or so there may have been only a few thousand individuals on the entire planet. This is just about as close to extinction as a species is likely to get and still bounce back, and—if true—must have placed our ancestors in as vulnerable a position as today's White Rhinos or Giant Pandas. Against all odds it seems that the dregs of our race managed to struggle through both the aftermath of Toba and the succeeding Ice Age, bringing our numbers up to the current 6 billion.

Could a future super-eruption wipe out the human race? It is highly unlikely that any eruption would be of sufficient size to completely obliterate today's teeming billions, but it is perfectly possible that our global technological society would not survive intact. Before the fall of the Berlin Wall, many national governments were quite prepared to plan for the terrible possibility of all-out nuclear war. With the threat now largely dissipated, however, there has been little enthusiasm for maintaining a civil defence plan to address the threat of a global geophysical catastrophe. In the absence of such forward thinking, the impact of a future super-eruption is likely to be appalling. With even developed countries such as the United States, the UK, Germany, and Australia having sufficient stores to feed their populations for a month or two at most, how would they cope with perhaps another six years without the possibility of replenishment? In the world's poorer countries, where famine and starvation are never far away, the situation would be magnified a thousand times, and death would come swiftly and terribly. From London to Lagos the law of the jungle would quite likely prevail as individuals and families fought for sustenance and survival. When the skies finally cleared and the Sun's initially feeble rays brought the first breath of warmth to the frozen Earth, maybe a quarter of the current population would have died through famine, disease, and civil strife.

Bearing in mind that over 70 millennia have elapsed since the Toba cataclysm it would be no surprise, statistically speaking, if another super-eruption struck within the next hundred years. But where? *Restless* calderas, which are constantly swelling and shaking, are clear candidates, and both Yellowstone and Toba belong in this category. Large volumes of magma still reside beneath these sleeping giants that may well be released in future cataclysms. It is likely, however, that the warning signs of these giants' awakening—large earthquakes and severe swelling of the surface—will continue for decades or even centuries before they finally let loose. As neither volcano is displaying such ominous behaviour at the moment we need not lose too much sleep over the imminence of a super-eruption at either Toba or Yellowstone. Only a tiny percentage of the Earth's 1,500 or so active volcanoes are currently, however, being monitored. Furthermore, the next super-eruption may blast itself to the surface at a point where no volcano currently exists. Perhaps even as

I write this some gigantic mass of magma that has been accumulating deep under the remote southern Andes may be priming itself to tear the crust apart—and our familiar world with it.

The super-eruptions I have talked about so far have all been cataclysmically explosive affairs. There is, however, another much less common species. One that—every few tens of millions of years—erupts even greater volumes of magma, but with relatively little violence. *Flood basalt* eruptions involve the effusion of gigantic volumes of low-viscosity lava that spread out over huge areas. These spectacular outpourings have been identified all over the world, including India, southern Africa, the northwest United States, and northwest Scotland, but the greatest breached the surface nearly 250 million years ago in northern Siberia. Estimates vary, but it looks as if the lavas erupted by this unprecedented event covered over 25 million square kilometres—an area three times that of the United States.

Several similar outpourings have occurred throughout the Earth's long history and have been correlated with mass extinctions. Before the Siberian outburst, for example, the Earth of the Permian period teemed with life. During the succeeding Triassic period, however, when the great flows had cooled and solidified, fully 95 per cent of all species had vanished from the face of the planet. A similar mass extinction 65 million years ago, at the end of the Cretaceous period, has been linked to the huge *Deccan Trap* flood basalt eruption in northwestern India. As I will address in the next chapter, however, there is incontrovertible evidence that the

Earth was struck at this time by a comet or asteroid, and many scientists now believe that this was the primary cause of the extinction of the dinosaurs and numerous other species at the end of the Cretaceous. Nevertheless, the Deccan lavas may also have had a role to play, pumping out gigantic quantities of carbon dioxide that may have led to severe greenhouse warming and the demise of organisms that were unable to adapt quickly enough. As our polluting society continues to do the same, perhaps we should take this as a salutary warning of what the future might hold for us, our world, and life upon it.

A watery grave

Although by no means the largest volcanic event of the twentieth century, the spectacular 1980 eruption of Mount St Helens, in Washington State (USA), was certainly the most filmed. Perhaps because it occurred in the world's most media-attuned country, the explosions as the volcano blew itself apart were almost drowned out by the whir of cameras and the scribbling of journalists' pencils. From a scientific point of view, however, the eruption was a watershed, because it drew attention to a style of eruption that had previously attracted little interest from volcanologists. Most eruptions involve the ejection of volcanic debris from a central vent, but the climactic eruption of Mount St Helens was quite different. Lava and debris from the previous eruption—all of 120 years earlier—had blocked the central conduit ensuring that the fresh magma rising into the volcano could not easily escape. Instead it forced its way into the volcano's north flank, causing it to swell like a giant carbuncle. By mid-May the carbuncle was 2 kilometres across and 100 metres high, and very unstable. Just after 8.30 in the morning on 18 May, a moderate earthquake beneath the volcano caused it to shrug off the bulge, which within seconds broke up and

25 The climatic eruption of Mount St. Helens (Washington State, USA) was triggered by the collapse of the north flank of the volcano

crashed down the flank of Mount St Helens as a gigantic landslide. With this huge weight removed from the underlying magma, the gases contained therein decompressed explosively, blasting northwards with sufficient force to flatten fully grown fir trees up to 20 kilkometres away and obliterating, in all, over 600 square kilometres of forest. The landslide material rapidly mixed with river and lake water forming raging mudflows that poured down the river valleys draining the volcano, while pyroclastic flows tore down the flanks and ash fell as far afield as Montana 1,000 kilometres away.

The Mount St Helens blast killed 57 people and was a disaster for the region, but its scientific importance lies squarely in its elucidation of the mechanism known as *volcano lateral collapse*. Most of us view volcanoes as static sentinels: bastions of strength and rigidity that are unmoving and unmovable. In fact, however, they are dynamic structures that are constantly shifting and changing. Far from being strong they are often rotten to the core; little more than unstable piles of ash and lava rubble looking for an excuse to fall apart. The numerous studies that followed the Mount St Helens eruption revealed that collapse of the flanks and the formation of giant landslides is a normal part of the lifecycle of many volcanoes, and probably occurs somewhere on the planet around half a dozen times a century. Furthermore, they showed that the Mount St Helens landslide was tiny compared to the greatest known volcano collapses—with a volume of less than a cubic kilometre compared with over 1,000 cubic kilometres for the prodigious chunks of rock that have, in prehistoric times, sloughed off the Hawaiian Island volcanoes.

At this stage you might be asking yourself, so what? Surely a hunk of rock—however large—falling off a volcano can't have a global impact—can it? Well it probably can, provided that the collapse occurs into the ocean. In 1792, a relatively small landslide flowed down the side of Japan's Unzen volcano and into the sea. The water displaced formed tsunamis tens of metres high that scoured the surrounding coastline, killing over 14,000 inhabitants in the small fishing villages that lined the shore. Just over a century later, in 1888, part of the Ritter Island volcano off the island of New Britain (Papua New Guinea) fell into the sea, generating tsunamis up to 15 metres high that crashed into settlements on neighbouring coastlines taking over 3,000 lives. Clearly, the combination of a volcanic landslide and a large mass of water is a lethal one, but—you are no doubt thinking—how can it affect the vast majority of the Earth's population who live far from an active volcano? The answer lies partly in the size of the largest collapses and partly in the scale of the tsunamis they generate.

Underwater images of the seabed surrounding the Hawaiian Islands shows that they are surrounded by huge aprons of debris shed from their volcanoes over tens of millions of years. Within this great jumbled mass of volcanic cast-offs, nearly 70 individual giant landslides have been identified, some with volumes in excess of 1,000 cubic kilometres. The last massive collapse in the Hawaiian Islands occurred just over 100,000 years ago from the flanks of the Mauna Loa volcano on the Big Island. Giant tsunamis resulting from entry of this huge mass of rock into the Pacific Ocean have been held responsible for carrying coral-reef debris to an altitude of over 300 *metres* above sea level on the neighbouring island of Lanai—three-quarters of the height of Chicago's Sears Tower. Deposits of a similar age, which may be tsunamis-related, have also been recognized 15 metres above sea level and 7,000 kilometres away on the southern coast of New South Wales in Australia. While the nature and provenance of both deposits is still debated, the scale of the waves generated appears to be realistic, and computer models developed to simulate the emplacement of giant volcanic landslides into the ocean come up with similar sized tsunamis.

It seems, then, as if major collapses at ocean island volcanoes are perfectly capable of producing waves that are locally hundreds of metres high and remain tens of metres high even when they hit land half an ocean away. The next collapse in the Hawaiian Islands is likely, therefore, to generate a series of giant tsunamis that will devastate the entire Pacific Rim, including many of the world's great cities in the United States, Canada, Japan, and China. In deep water, tsunamis travel with velocities comparable to a Jumbo Jet, so barely 12 hours will elapse before the towering waves crash with the force of countless atomic bombs onto the coastlines of North America and eastern Asia.

Nor is the problem confined solely to the Pacific. Scientific cruises around the Canary Islands, together with detailed geological surveys on land, have revealed a picture very similar to that painted for Hawaii. Huge masses of jumbled rock stretching for hundreds of kilometres across the seabed, and gigantic cliff-bounded collapse scars on land, testify to enormous prehistoric collapses from the islands of Tenerife and El Hierro. Of more immediate concern, it looks as if a new giant landslide has recently become activated on the westernmost Canary Island of La Palma, and is primed and ready to go. During the eruption before last, in 1949, much of the western flank of the island's steep and rapidly growing volcano—the Cumbre Vieja—dropped 4 metres towards the North Atlantic and then stopped. Some UK and US scientists believe that this gigantic chunk of volcanic rock—with an estimated volume of a few hundred cubic kilometres, just about double the size of the UK's Isle of Man—is now detached from the main body of the volcano and will eventually crash *en masse* into the sea. The problem at the moment is that we don't have a clue when this will happen. It will probably be soon—geologically speaking—but whether it will be next year or in 10,000 years we simply don't know. Measurements undertaken during the late 1990s using the satellite Global Positioning System proved somewhat inconclusive but suggested that the landslide might still be creeping slowly seawards, perhaps at only a centimetre a year or even less. Even if this is the case, however, the rock mass is unlikely to complete its journey into the North Atlantic without the trigger of a new eruption.

What is certain is that at some point in the future the west flank of the Cumbre Vieja on La Palma *will* collapse, and the resulting tsunamis will ravage the entire Atlantic rim. Steven Ward of the University of California at Santa Cruz and Simon

26 The west flank of La Palma's Cumbre Vieja volcano is on the move and will eventually collapse into the North Atlantic

Day of University College London's Benfield Greig Hazard Research Centre created quite a stir recently when they published a scientific paper that modelled the future collapse of the Cumbre Vieja and the passage of the resulting tsunamis across the Atlantic. Within two minutes of the landslide entering the sea, Ward and Day show, an initial dome of water an almost unbelievable 900 metres high will have been generated, although its height will rapidly diminish. Over the next 45 minutes a series of gigantic waves up to 100 metres high will pound the shores of the Canary Islands, obliterating the densely inhabited coastal strips, before crashing onto the African mainland. As the waves head further north they

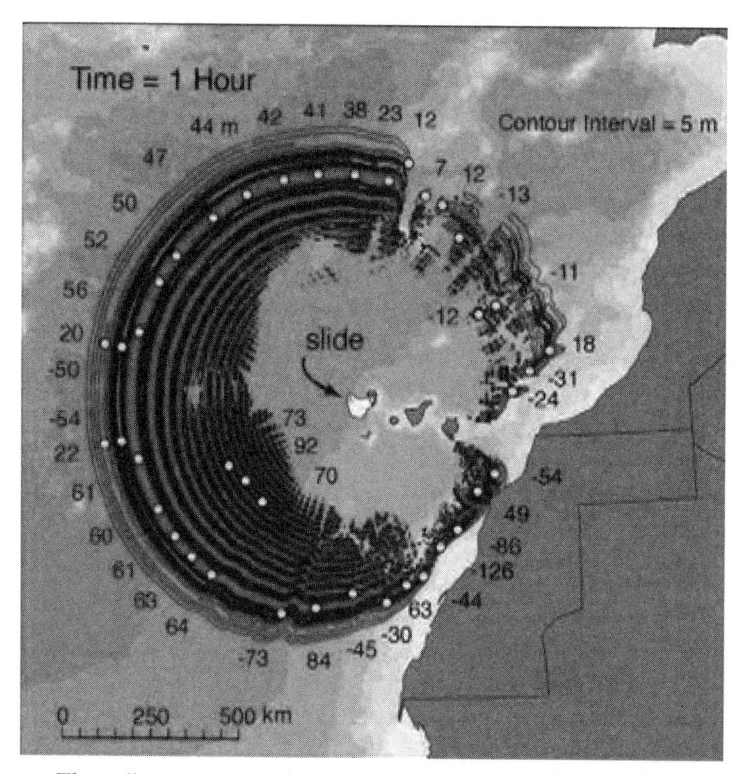

27 The collapse of the Cumbre Vieja will generate enormous tsunamis that will batter Africa, Europe, and the Americas. Wave-crest heights are indicated by positive numbers and troughs by negative numbers (all heights in metres). The map shows the position of the waves one hour after collapse

will start to break down, but Spain and the UK will still be battered by tsunamis up to 7 metres high. Meanwhile, to the west of La Palma, a great train of prodigious waves will streak towards the Americas. Barely six hours after the landslide, waves tens of metres high will inundate the north coast of Brazil, and a few hours later pour across the low-lying islands of the Caribbean and impact all down the east coast of the United States. Focusing effects in bays, estuaries, and harbours may increase wave heights to 50 metres or more as Boston, New York, Baltimore, Washington, and Miami bear the full brunt of Vulcan and Neptune's combined assault. The destructive power of these skyscraper-high waves cannot be underestimated. Unlike the wind-driven waves that crash every day onto beaches around the world, and which have wavelengths (wave crest to wave crest) of a few tens of metres, tsunamis have wavelengths that are typically hundreds of *kilometres* long. This means that once a tsunami hits the coast as a towering, solid wall of water, it just keeps coming—perhaps for ten or fifteen minutes or more—before taking the same length of time to withdraw. Under such a terrible onslaught all life and all but the most sturdily built structures will be obliterated.

Without considerable forward planning it is unlikely that the nine hours it will take for the waves to reach the North American coastline will permit effective, large-scale evacuation, and the death toll is certain to run into millions if not tens of millions. Furthermore, the impact on the US economy will be close to terminal, with the insurance industry wiped out at a stroke and global economic meltdown following swiftly on its heels. In this way, a relatively minor geophysical event at a remote Atlantic volcano will affect everyone on the planet. Like volcanic super-eruptions, these giant tsunamis constitute perfectly normal, albeit infrequent, natural phenomena. At some point in the future one will certainly wreak havoc in the Atlantic or Pacific Basins, but when? The frequency of collapses on the Hawaiian volcanoes has variously been estimated to be between 25,000 and 100,000 years, but if giant landslides at all volcanic islands are considered, it may be that a major collapse event occurs every ten millennia or so. On a geological timescale this is very frequent indeed and should provide us with serious cause for concern. Even more worryingly, the rate of collapse may not be constant and the current episode of global warming engendered by human activities may in fact bring forward the timing of the next collapse. My own research team has linked increased incidences of past volcano collapse with periods of changing sea level, while others have suggested that a warmer and wetter climate might result in greater numbers of large volcanic landslides. Given that sea levels are forecast to continue to rise for the foreseeable future, while studies of past climate change show that a warmer planet results in heavier rainfall on many of the world's largest volcanic island chains, perhaps we should all be thinking of moving inland and uphill, or at least of investing in a good-quality wet suit.

The city waiting to die

It is extraordinarily difficult to get across to someone who has never experienced it the sheer, mind-numbing terror of being caught in a major earthquake. Even in California, where the population is constantly bombarded with information about what to do in the event of a quake, coherent, sensible thought ceases when the ground starts to tremble. Following the Loma Prieta quake that struck northern California in 1989, a survey by the United States Geological Survey revealed that only 13 per cent of the population of Santa Cruz sought immediate protection, while close to 70 per cent either froze or ran outside. This is a perennial problem with earthquakes; however well educated the people, when the ground starts bucking like a bronco and the furniture starts to hurl itself across the room, blind instinct takes over and tells them to 'get the hell out of there'. Unfortunately, this serves only to increase the death toll as terrified homeowners rushing screaming into the street provide easy targets for falling masonry and other debris crashing down from above. What they should be doing is diving beneath the nearest piece of heavy furniture or sheltering beneath the lintel of a convenient doorway.

Earthquakes are immensely destructive, mainly because most cities in regions of high seismic risk are dominated by buildings that are simply not built well enough to withstand the severe ground shaking of a major quake. Modern construction methods in California follow stringent building codes that ensure they can withstand earthquakes that would be devastating elsewhere, and this policy has borne considerable fruit by dramatically limiting death, injury, and damage during major quakes in the last 15 years. Even so, the Northridge earthquake that struck southern California in 1994 is credited with losses totalling 35 billion US$, largely accruing from damage to older structures. Other earthquake-prone countries also have in place building codes designed to minimize damage due to ground shaking, but often these codes are simply not enforced. The terrible legacy of such a lack of commitment by government and local authorities became all too apparent when a magnitude 7.4 quake struck the Izmit region of Turkey in 1999, obliterating 150,000 buildings and taking over 17,000 lives. Many apartment blocks simply *pancaked*; successive floors collapsing to form a pile of concrete slabs beneath which opportunities for survival were minimal. More recently, in January 2001, a severe earthquake shook the Bhuj region of Gujarat state in northwestern India, flattening 400,000 homes and killing perhaps 100,000 people. Many of the deaths resulted from the traditional construction methods used in the region, which involved the building of homes with enormously thick walls made of great boulders held together loosely with mud or cement, beneath heavy stone roofs. When the ground started to shake these buildings offered little resistance, collapsing readily to crush those inside.

During the last millennium, earthquakes were responsible for the deaths of at least 8 million people. Terrifying as this sounds, the rapid growth of megacities in regions of high seismic risk is set to ensure that this figure is surpassed, maybe in just the next few centuries, and some seismologists are already warning of the potential, in the near future, for a single large quake to take 3 million lives. If the unfortunate target were Karachi or Mexico City then, although the catastrophe would have appalling consequences for the host countries, the global impact would be minimal and would barely impinge upon the lives of most of the world's population. On the other hand, if ground zero were to be the Japanese capital, Tokyo, then the story would be very different. Projections to 2015 suggest that by this time the Tokyo-Yokohama conurbation will be the greatest urban concentration on the planet, with a population a shade under 29

28 Buildings, not earthquakes, kill people. 400,000 poorly constructed buildings succumbed to severe ground shaking during the 2001 Gujarat earthquake in India

million. The city is located in one of the most quake-prone parts of the planet, where the Pacific and Philippine plates to the east plunge beneath the giant Eurasian plate, and was obliterated by a massive earthquake less than 80 years ago. While things have been ominously quiet since, it can't be long now before another huge quake devastates one of the world's great industrial powerhouses. When it does, the economic shock waves will hurtle out across the planet, bringing country after country to its knees. In order to provide an impression of the fate awaiting the Japanese capital, let me take you on a trip back to one of the great disasters of the twentieth century, the terrible event the Japanese call the Great Kanto Earthquake.

September 1st, 1923, dawned like any other day for the inhabitants of Tokyo and Yokohama, but for many it would be their last. The quake struck just before noon, when the cafés and beer halls were packed with hungry workers and as families sat down at home to their midday meal. A low, deep rumbling grew rapidly to a monstrous roar as a fault below Sagami Bay ripped itself apart and sent shock waves tearing northwards towards the twin cities, crashing first into Yokohama and then—a bare 40 seconds later—into the heart of the capital itself. The quake registered a massive 8.3 on the Richter Scale, and so severe was the ground shaking that it was impossible even to stand. Within seconds, thousands of buildings, many with the traditional wooden walls and heavy tiled roofs, collapsed into heaps of rubble, bringing sudden oblivion to those inside. The great cacophony of grinding rock and falling buildings eventually gave way to the quieter

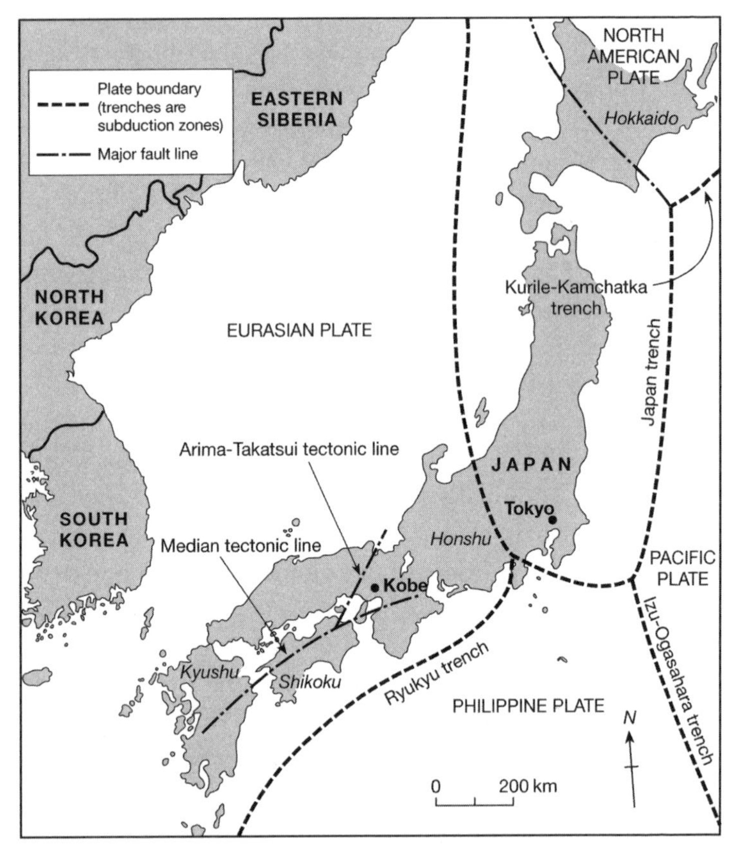

29 The Japanese capital sits in a region of complex geology where three of the planet's great tectonic plates meet

but equally terrifying crackling of flames as fires started by thousands of overturned stoves began to devour the wood out of which many of the buildings were constructed. Whipped up by a brisk wind, a million small fires swiftly merged to form unstoppable walls of flame that marched across the ruins. Shocked men, women, and children cowered before them in open spaces, but to no avail. The firestorms roasted them alive. In one area of waste ground 40,000 were immolated by the conflagration, so packed together that their charcoaled bodies were found still upright. The fires continued to consume what remained of the cities for two days and nights, before finally burning themselves out to reveal a post-apocalyptic scene of utter devastation. The true total will never be known but up to 200,000 people may have lost their lives in the quake itself and the fires that followed. The cost to the Japanese economy was phenomenal—around 50 billion US$ at today's prices—and a combination of the quake and the Great Depression six years later led to economic collapse and severe hardship. Some have even suggested that these circumstances, as in the German Weimar Republic, helped stoke the fires of nationalism and the rise of the military, leading to conquest, imperialism, and ultimately war.

In the early years of the new millennium, the twin cities of Tokyo and Yokohama again await their fate, only this time it will be far, far worse, both for Japan and the rest of the world. Now the industrial and commercial might of the region constitutes one of the major hubs of the world market, with spokes reaching out to the far corners of the Earth, helping

30 Little remained standing after the post-quake fires had raged for two days across Tokyo following the Great Kanto Earthquake of 1923

to bind together a global economic machine upon which the wealth of all nations now depends. When Tokyo falls, so will Japan, and the rest will follow—but when? Strains have now been accumulating in the rock beneath and around the capital for 78 years and, apart from a relatively small—magnitude 5.9—quake in 1992, the region has been seismically silent. Both the government and the population know, however, that this can't last and money is being poured into constructing earthquake-proof buildings, improving education and emergency planning, and even trying to predict the precise timing of the next 'big one'. So far, however, the accurate prediction of earthquakes has proved to be out of reach, and prospects for a breakthrough in the near future are slim. Furthermore, a substantial proportion of the older building stock remains vulnerable, and an estimated one million wooden buildings continue to provide an excellent potential source of fuel for the post-quake fires. Just seven years ago, 6,000 people died in the Kobe earthquake, 400 kilometres south of Tokyo, which can be viewed perhaps as a mini version of the catastrophe awaiting the capital. At Kobe serious fire damage contributed significantly to the overall destruction and to the huge economic losses of 200 billion US$, and it was clear that emergency preparedness and response were far from effective, and certainly well below the rest of the world's expectations, given the general perception of Japanese society as a model of efficiency. For one reason or another, the authorities were simply unable to cope with the chaotic aftermath of the event. Plans were not in place to ensure transport of emergency supplies and equipment to where they were needed, once roads were blocked by debris and railways out of commission, and many of the city's hundreds of thousands of homeless received little or no help for several days after the quake. It is fair to say that some at least of the problems encountered at Kobe reflect the hierarchical structure of Japanese society, which stifles independent decision making and action and hinders rapid response in emergency situations. Without significant changes it is difficult to see how any earthquake emergency plan for the Tokyo region could function effectively within the strait-jacket imposed by such a deeply ingrained and restrictive social etiquette.

The geological setting of Tokyo and Yokohama is complex, with three of the Earth's great tectonic plates converging here. The enormous strains associated with the relative movements of these plates are periodically relieved by sudden displacements along local faults, which in turn lead to destructive earthquakes. In fact, there are so many active faults in the vicinity that the region is at risk from major quakes occurring at four different locations, all of which are thought by seismologists to be overdue or at least imminent. Some 75 kilometres south of Tokyo and Yokohama, close to the city of Odawara, seismologists expect a magnitude 6.5–7 quake to strike at any moment. Although causing serious damage locally, and moderate damage in the twin cities, this is unlikely to hit the capital with the force of the 1923 quake. Similarly, another so-called *Tokai* earthquake is imminent beneath Suruga Bay, 150 kilometres to the southwest. Scientists forecast that this will be a huge, magnitude 8 event that will undoubtedly batter the coastal city of Shozuoka, but will probably again be too far from the capital to have a serious impact. Far more worrying are two other expected quakes that pose a much greater threat to the Tokyo region, and which are awaited with much trepidation. Scientists predict that a quake as large as magnitude 7 could strike at any time—directly beneath the capital. This event, known locally as a *chokka-gata* quake, will cause severe damage in the capital, although Yokohama is likely to be less badly hit. Worst of all, a repeat of the 1923 Great Kanto Earthquake itself is only thought to be a few decades away. This is likely to take the form of a massive magnitude 8 event resulting from the tearing open of a fault beneath Sagami Bay to the south. As was the case almost 80 years ago, the shock waves will race northwards, rolling first into Yokohama and barely half a minute later into Tokyo itself.

The national government still maintains that its scientists will detect in advance the warning signs that the 'big one' is on its way. Such faith in science is both rare and touching but in this case entirely misplaced. Retrospectively, it has been noted that some earthquakes have been preceded by falls in the water levels in wells and boreholes, and in elevated concentrations of radioactive radon gas issuing from the rock, but this is not always observed. Furthermore, such changes can occur without a following quake, making them notoriously unreliable for prediction purposes. A group of Greek scientists claim that they can detect electrical signals in the crust prior to an earthquake, but there is no convincing evidence for this and the method is derided by most seismologists. On the other hand, there does appear to be something in the idea that animals, birds, and fish behave strangely before a quake, and the Japanese are actually undertaking serious research to find out if catfish—amongst other organisms—can help them forecast the next big one. The problem here is that no one knows how animals can detect a quake before it happens, although it has been suggested that strain in the rocks generates electrical charges in fur and feathers, and perhaps even scales, that trigger small electric shocks, making the animals understandably restless and irritable. But this begs the question, how *do* you decide whether, for example, a pig is behaving strangely?

In the absence of an alert from a precognizant catfish, it is likely then, that the next great quake will strike the Tokyo region with no warning whatsoever. Recently constructed buildings will fare reasonably well, but many older properties will crumble. Notwithstanding an automatic gas shut-off device that is fitted to some buildings, exploding fuel tanks, fractured gas mains, and oil and chemical spills will ensure no shortage of fires to feed on a million wooden buildings. As in 1923, huge conflagrations are expected to cause at least as much destruction as the quake itself, and to inflate the death toll hugely, which some suggest could easily top the 200,000 of the Great Kanto quake. While it is difficult to estimate in advance the economic losses resulting from the next big one, a modelling company that services the insurance industry has come up with the extraordinary figure of 7 *trillion* US$. This would make the cost of the next Tokyo quake 35 times greater than Kobe, so far the most expensive natural disaster ever, and 200 times more than California's 1994 Northridge quake—the costliest natural catastrophe in US history.

The impact on the Japanese economy is widely expected to be shattering. Japan is enormously centralized, and the Tokyo region hosts not only the national government but also the stock market and 70 per cent of the headquarters of the country's—and the world's—largest companies. Currently, the GDP of the region is comparable to that of the entire UK, and despite its current economic woes, it is likely that this will be substantially larger when the big one eventually strikes. In order to rebuild and regenerate it is highly likely that the Japanese will have to disinvest from abroad on a massive scale, dumping government bonds in Europe and the States, selling foreign assets, and shutting down overseas factories. It is well within the realms of possibility that as country after country finds itself fighting to cope with the swift unravelling of the global economy, a recession deeper than anything since the 1920s would soon set in. Neither would it be any great surprise to find unemployment reaching staggering proportions and the political and social fabric of many states starting to pull apart. No one knows how long a post-Tokyo quake depression would last—it could be years or even decades—nor just how bad it would be. Equally importantly, how long do we have to wait until such a speculative scenario is played out for real? The frightening consensus amongst seismologists is 30 years—at best.

Despite occasionally being depicted in the media as 'Disasterman', I would hate you to regard me purely as a harbinger of doom, and close the book at this point with a feeling of hopelessness about the future. Yes, the Earth is geologically very dangerous, and the more geologists study our planet the more potentially serious the tectonic threat to the survival of our civilization appears to be. On the other hand, we are learning all the time; collecting data that can be utilized to counter or at least mitigate the impact of the next super-eruption or gigantic tsunamis. Eventually, it probably will be possible to predict earthquakes with some accuracy and precision, and certainly within a century it is likely that nowhere on the planet will a volcanic island become unstable or a huge new batch of magma swell the surface without our satellites spotting them well in advance of catastrophe. On an almost daily basis Earth scientists are tackling some of the greatest threats to our society and incrementally they are getting to grips with them. At the very least, the next time our planet shudders on a grand scale we will be far better prepared than our distant ancestors, who faced the might of Toba with incomprehension and sheer terror.

Facts to fret over

- On average there are two volcanic super-eruptions every 100 millennia.
- Following the Toba super-eruption 73,500 years ago, the world would have been held in the grip of volcanic winter for at least six years.
- In the aftermath of Toba the human population may have been reduced to just a few thousand individuals.

- In 1949 a gigantic landslide on the western flank of the Cumbre Vieja volcano on La Palma (Canary Islands) dropped 4 metres over night.
- When the Cumbre Vieja collapses into the sea, the coastal cities of the eastern USA will be battered by tsunamis up to 50 metres high.
- The next great Tokyo earthquake is expected to cause damage totalling at least 7 trillion US$ and may trigger a global economic collapse.

The Threat from Space

Asteroid and Comet Impacts

The astronomical event of the century

In 1993 a discovery by the late and greatly lamented planetary scientist Eugene Shoemaker—along with his wife Carolyn and colleague David Levy—was to change for ever our perception of the Earth as a safe and cosy haven insulated from the whizzes and bangs of a violent and capricious universe. The Shoemaker team had spotted 21 huge chunks of rock that had once been part of a comet torn apart by the enormous gravitational field of the planet Jupiter—a giant sphere mainly made up of hydrogen and helium gas that is large enough to contain over 1,300 Earths. Instead of orbiting the Sun, like most comets, however, this one had been captured by Jupiter's gravity and the rocky fragments now orbited the King of Planets itself. As Jupiter already had a large retinue of moons, the addition of a few more would have been mildly interesting, if not surprising. What was extraordinary, however, was that these new 'moons' were very much ephemeral. The following year they would end their lives by crashing into the surface of Jupiter, providing scientists on Earth with a grandstand view of just what happens when a planet is struck by large hunks of space debris.

On 16 July 1994—appropriately the 25 anniversary of the launch of *Apollo 11*, the first manned lunar landing mission—the first fragment of Comet Shoemaker-Levy struck the planet, sending up a gigantic plume of gas and debris and blasting outwards a rapidly expanding shock wave. As fragment after fragment hammered into the planet, spectacular images were gathered by the Hubble Space Telescope in Earth orbit and by the unmanned *Galileo* probe on its way to Jupiter. Two days after the initial impact, a chunk of rock 4 kilometres across and rather unromantically named fragment G smashed into the planet with the force of 100 million million tonnes of TNT—roughly the equivalent of eight *billion* Hiroshima-sized atomic bombs. The flash generated by the collision was so brilliant that many infra-red telescopes trained on the event were temporarily blinded. The glare soon faded, however, to reveal an enormous dark impact scar wider than the Earth. Inevitably, everyone who saw this awesome image had the same thought. What would have happened if fragment G had struck the Earth instead of Jupiter?

Almost overnight our planet seemed a much more vulnerable place and the hold of our race upon it that much more tenuous. Suddenly both scientists and the public, and even politicians, began to take the threat from space seriously. Two Hollywood blockbuster films fed growing interest in impact events by showing—with various degrees of scientific rigour—what we might all be in for if a comet or asteroid headed our way.

In 1996, just two years after the Jupiter impacts, an international body known as the *Spaceguard Foundation* was formed, with the dedicated aims of promoting the search for potentially dangerous asteroids and comets and raising the general level of awareness of the impact threat. In the United

States, NASA and the Department of Defence began, albeit at a low level, to fund Spaceguard-related projects and the UK government established a task force to examine the risk of asteroids and comets hitting the Earth. All of a sudden everyone wanted to know what the chances *were* of the Earth being

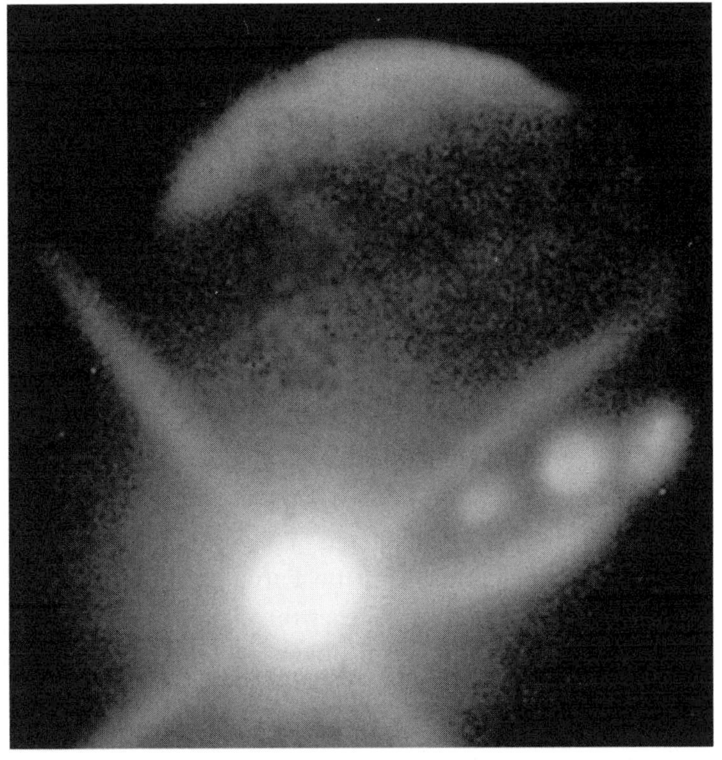

31 The spectacular impact of one of the fragments of Comet Shoemaker-Levy impacts on Jupiter in 1994

struck at some point in the future and what effect such a collision would have on our planet and our race. The answer to the first question is easy: the chances are 100 per cent. The Earth has been bombarded by space debris throughout its long history, and although such collisions are now far less common than they were billions of years ago, our planet will be struck again. The vital question is—when? And as regards how bad this will be for the human race: that depends largely upon how big a chunk of rock hits us.

The cosmic sandstorm

To get a better idea of how frequently the Earth is likely to be hit, we need to find out how many rocks are hurtling around our solar system and, in particular, how many of these come close enough to the Earth to start us worrying. Although a vast amount of debris was swept up by the embryonic planets during the early solar system, countless dregs remain, ranging upwards in size from tiny specks a few millimetres across to gigantic rocks, such as the minor planet Ceres, over a 1,000 kilometres in diameter. Like someone battling through a desert sandstorm, the Earth is constantly bombarded in the course of its journey through the solar system. Fortunately for us, most of the billions of colliding fragments are tiny and flash into oblivion as soon as they come into contact with our planet's atmospheric shield. Every now and again, however, the Earth collides with something larger.

A fragment of debris the size of a pea burns up in the

Earth's atmosphere every five minutes, while a soccer-ball sized lump will light up the sky with its death throes around once a month. Larger objects may run the gauntlet of the atmosphere and reach the surface, but this is rare and only happens a few times a year. Perhaps once or twice a century, the Earth collides with a rock in the 40–50 metre size range—an object large enough to obliterate a city if it scores a direct hit. The last well-documented impact of this size occurred as recently as 1908—of which more later.

While the entire solar system teems with debris, from a hazard point of view we are only really interested in those fragments that threaten to end their existence through collision with our planet. The majority of these Earth-threatening objects are rocky *asteroids* that have orbits around the Sun that intersect the Earth's. The true numbers of these *Earth-Crossing Asteroids* (ECAs) are impossible to determine, but current estimates are pretty frightening.

In all, up to 20 million pieces of rock over 10 metres across may be hurtling across our planet's path during its journey around the Sun. Up to 100,000 of these are thought to be over 100 metres in diameter—big enough to obliterate London or New York given a direct hit—and maybe 20,000 are half a kilometre across, sufficient to wipe out a small country if they strike land, or generate devastating tsunamis if they impact in the ocean. Fewer in number, but enormously more destructive if they hit, are those asteroids 1 kilometre or more in diameter. One kilometre is widely recognized as the critical size threshold at or above which a collision would have devastating consequences on a global scale. Although barely equivalent in diameter to ten soccer pitches laid end to end, such is the tremendous level of kinetic energy—or energy of motion—involved in the collision that an object of this size striking land would leave a crater 20 kilometres or so across and loft sufficient pulverized debris into the atmosphere to block out the Sun's rays and plunge the Earth into a freezing *cosmic winter* for years.

A range of estimates have been published for the number of Earth-Crossing Asteroids in the 1 kilometre and above size range, with the most recent suggesting there are somewhere between 500 and 1,100. Over 320 of these have now been identified and their orbits projected forward in time to see if they pose a threat to the Earth in the medium term, and the search continues to find them all—a task that will take at least a couple more decades. Once this has been accomplished, and assuming that one does not have our name on it, we can sleep a little safer in our beds. The problem does not, unfortunately, end there. We still have the comets to worry about.

Comets are enormous masses of rock and ice that can be up to 100 kilometres or more across. In contrast to the near-circular orbits of the asteroids, most comets follow strongly elliptical paths that carry them from the freezing outposts of the outer solar system, or beyond, in close to the Sun and then out again. In the depths of space, comets are enigmatic objects and not easy to spot. As they enter the inner solar system, however, they undergo a remarkable transformation as sunlight starts to evaporate gas and dust particles from the central *nucleus* to form a spectacular *tail* that can stretch

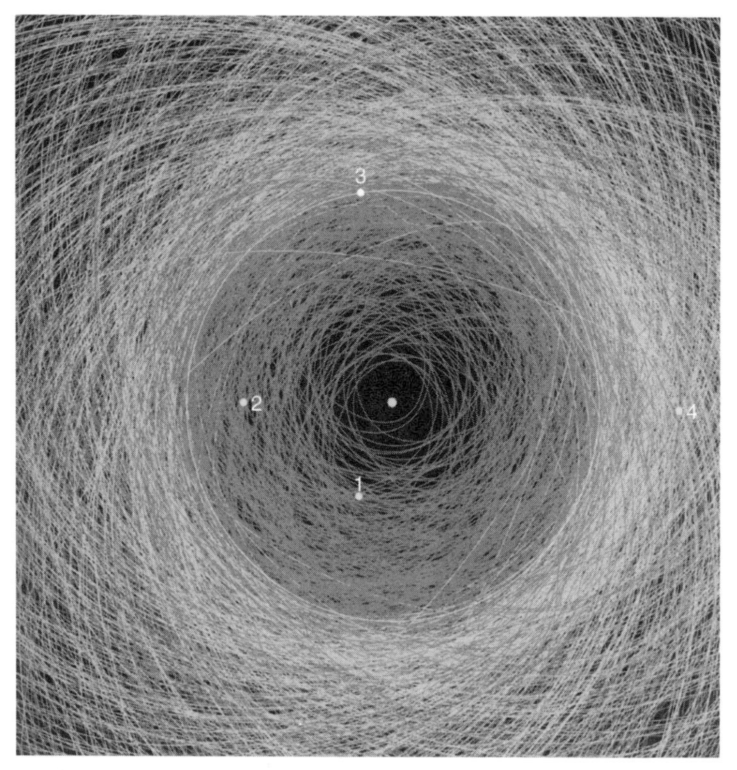

32 Orbits of known Earth-crossing Asteroids provide a clear picture of just how crowded the space around our planet really is. The orbits of 1. Mercury 2. Venus 3. Earth 4. Mars are also shown across space for 100 million kilometres or more. The stunning *apparition* of a comet's tail was long regarded as a portent of doom and disaster, and in a way this is not too far from the mark. Comets have typical velocities of 60–70 kilometres a second—a hundred times faster than Concorde, and around three times that of the Earth-Crossing Asteroids. As a result, a collision between a comet and the Earth is hugely more energetic and therefore tremendously more destructive.

Another problem with comets is that, unlike their asteroid cousins, their orbital parameters are often poorly known and therefore difficult to project into the future to see if they pose any threat. Halley's Comet, undoubtedly the most famous of all, follows an orbit around the Sun that takes only 76 years to complete. Consequently, it has been observed dozens of times over thousands of years and its orbit is well enough known to make it possible to calculate its path far into the future. This shows that, at least until 3000 AD, Halley's Comet will not even come close to threatening the Earth. Other comets, however, follow *parabolic* orbits that take them on immeasurably long journeys far beyond the limits of the solar system. Some of these may have been observed once or twice by our distant ancestors, but others may be making their first ever visit to the inner solar system. In these circumstances, there has been no opportunity to predict their orbits on the basis of earlier visitations, and our first view of one of these objects heading our way may provide us with just six months respite before an unavoidable and calamitous collision. Furthermore, because such comets have been confined to deep space, they are huge—perhaps 100 kilometres or more across. This is because they have not suffered attrition from the *solar wind,* the hurricane of solar particles that evaporates parts of a comet to produce the characteristic tail, as it forges its way through the inner solar system.

When worlds collide

It has been something of a struggle over the last few centuries for exponents of the theory to convince both scientists and the public that the millions of craters that pockmark the face of the Moon are the result not of volcanic explosions but of collisions with objects from space. As long ago as the early nineteenth century, the German natural philosopher Baron Franz von Paula Gruithuisen's declaration that the lunar craters were a consequence of 'a cosmic bombardment in past ages' was treated with contempt by 'serious' scientists. (No doubt his further claims to have uncovered evidence for the existence of humans and animals on the surface of the Moon had a little to do with this.) At the other end of the nineteenth century, the US geologist Grove Karl Gilbert tried to simulate in the laboratory the formation of the lunar craters by firing objects into powder or mud. Gilbert was perplexed, however, by the observation that only objects fired vertically produced circular craters like those that cover the lunar surface. In light of this, W. M. Smart proclaimed, in 1927, that the craters of the Moon could not be caused by impacts because '*there is no a priori reason why meteors should always fall vertically*'. It was only after observing the effects of the billions of tonnes of bombs dropped in the Second World War that it began to dawn on geologists that given a violent enough explosion, a circular crater would always be formed—whatever the angle of the trajectory. In other words, the tremendous explosion generated when an object hit the Moon virtually always resulted in a circular crater. Remarkably, it took another quarter of a century for the impact origin of lunar craters to gain widespread acceptance, and even today one or two maverick scientists still support a volcanic origin in the face of overpowering evidence to the contrary. Getting any new paradigm accepted in science is a battle, and geology is no exception. Just as the proponents of the revolutionary theory of plate tectonics had initially to fight hard against reactionary forces, so those scientists who claimed that the Earth, as well as the Moon, had also taken a battering found the going difficult.

As long ago as 1905, Benjamin Tilghman proposed that Arizona's famous Barringer Crater (also now known as Meteor Crater) was the result of 'the impact of a meteor of enormous and hitherto unprecedented size'. This suggestion failed to convince, however, because a quarter of a century of excavation by Tilghman and his engineer colleague D. M. Barringer failed to find the impactor itself. We now know that this had been essentially vaporized by the enormous heat generated by the collision, but at the time the absence of a 'smoking gun' simply lent credence to those who suggested an alternative mechanism of formation.

Until well after the end of the Second World War, many Earth scientists suffered an extraordinary failure of the imagination, accepting an impact origin for the lunar craters but grabbing at any straw in order not to support an impact origin for crater structures on the surface of our own planet. Given that, due to the Earth's much greater size and stronger gravity field, it must have been struck perhaps 30 times more frequently than its nearest neighbour, this denial is even more extraordinary. Perhaps not entirely surprising, however, when we consider that the enormously dynamic nature of our planet is far from suited to the preservation of impact craters, particularly those of any great age. Because of plate tectonics, and in particular the process of subduction, through which the basaltic oceanic plates are continuously being consumed within the Earth's hot interior, some two-thirds of the Earth's surface is only a few hundred million years old. Bearing in mind that the most intense phase of bombardment occurred during the first few billion years of our planet's history, then evidence for this can now only be found in the ancient hearts of the granite continents that are immune to the subduction process. Because they have succumbed to erosion and weathering, perhaps for aeons, these craters are notoriously difficult to spot. Also, the oldest rocks, which are likely to support the most craters, are in remote areas such as Siberia, northern Canada, and Australia, and some craters are so big that their true form can only be seen from space. Today, satellites have helped in the identification of over 165 impact craters all over the world, and the idea that the Earth is susceptible to bombardment from space is now as accepted as plate tectonics.

Controversy has certainly not gone away, however, and argument continues amongst the scientific community, particularly about the frequency and regularity of impacts and—probably of most interest to the layman—about the effects of the next large impact on our civilization. The question of frequency is far from straightforward and serious disagreement exists between schools of thought that, on the one hand, support a constant flux of impactors and, on the other, advocate so-called *impact clustering*. Notwithstanding the very heavy bombardment of the Earth's early history, followers of impact *uniformitarianism* support a strike rate that is uniform and invariable. This is at variance with rival groups of scientists who are promoting an alternative theory of *coherent catastrophism*, within which the Earth, for one reason or another, periodically comes under attack from increased numbers of asteroids or comets.

If we are realistically to assess the threat of future impacts to our civilization, then clearly it is vital that we resolve as soon as we can whether the number of collisions continues at its current rate or whether we have a nasty shock in store somewhere down the line. If the former proves to be correct, we can expect business as usual, meaning a collision with a 50-metre potential city-destroyer every 50 years or so, a half-kilometre small-country obliterator every ten millennia, and a 1-kilometre global impact event every 100,000 to 333,000 years—depending on whose figures you accept. Fortunately for us, gigantic *extinction level events* (ELEs), such as that

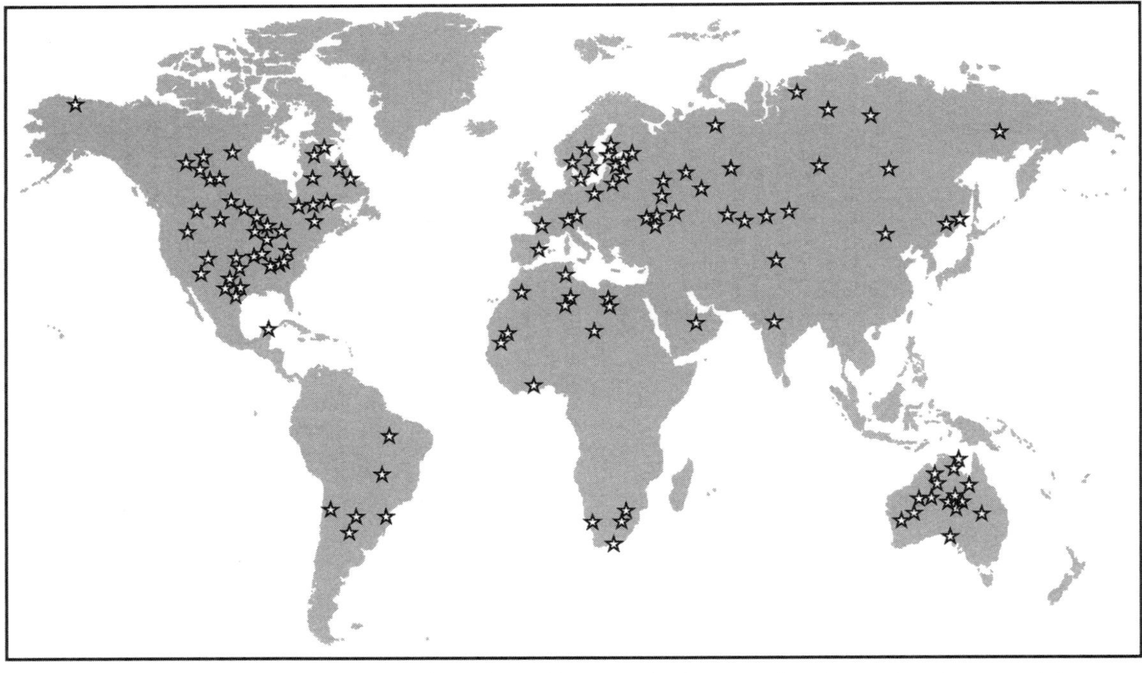

33 Over 165 impact craters have now been identified on Earth, many in the ancient hearts of the continents

caused by the 10-kilometre monster that ended the reign of the dinosaurs 65 million years ago, appear to happen every 50 to 100 million years, so the chances of one striking the Earth soon are tiny. Based upon the above impactor strike rates, proponents of the threat from asteroids and comets come up with probabilities of dying due to an impact that really make one think. If you were able to construct a time machine and hurtled forwards to the year 1,000,2002 where you sought out and consulted the Centre for Planetary Records you would come up with a fascinating fact. The number of people killed in air (and no doubt space) crashes during the intervening million years—probably between 1 and 1.5 billion—would be less than those killed by impact events, which could total 3 billion or more, assuming two or three collisions with 1-kilometre objects. What this amounts to is that during your lifetime your chance of dying due to an asteroid or comet impact could be twice as great as being killed in an air crash; a pretty sobering thought if ever there was one. Looked at another way, if you gamble, your chance of being killed during an asteroid or comet strike is 750 times more likely than winning the UK lottery. Maybe this scares the wits out of you, but the true situation may actually be worse. If the coherent catastrophists are correct then there are certain periods in the Earth's history when our planet, or perhaps even our entire solar system, travels through a region of space containing substantially more debris than normal, resulting in a significant increase in impact events on all scales.

A number of theories lay the blame for this periodic

increase in Earth-threatening space debris on the episodic disruption of the so-called *Oort Cloud,* a great spherical cloud of comets that envelops the entire solar system far beyond the orbit of Pluto. Typically, comets in the cloud travel along such huge orbits, which take some a quarter of the way to the nearest star, that they rarely visit the inner solar system, and then only in ones and twos. However, if some external influence were to interfere with the cloud, so the thinking goes, hundreds or thousands could have their orbits changed encouraging them to plunge Sunwards, greatly raising the threat of collision with the planets—including our own. A number of suggestions have been put forward for how the

34 The Barringer Crater (also known as Meteor Crater) in Arizona is the legacy of a collision with a small asteroid 50,000 years ago

Oort Cloud might be periodically disrupted, including due to the passage through the cloud of the mythical and much sought after *planet X*, which some scientists think may be orbiting far beyond frozen Pluto, or to a dark and distant stellar companion of our own Sun.

An alternative and intriguing theory, known as the *Shiva hypothesis* after the Hindu god of destruction and renewal, has been vigorously promoted by Mike Rampino of New York University and his colleagues, who believe that the great extinctions recognized in the Earth's geological record are the result of major impact events that happen pretty regularly every 26–30 million years. Rampino and his colleagues link this to the orbit of our solar system—including the Earth—about the centre of our Milky Way galaxy, an orbit that moves up and down in a wave-like motion. Every 30 million years or so, this undulating path takes the Sun and its offspring through the plane of our disc-like galaxy, when the gravitational pull from the huge mass of stars at the galaxy's core provides an extra tug. This, say the Rampino school, is sufficient to disturb the orbits of the Oort Cloud comets to an extent sufficient to send an influx of new comets into the heart of the solar system, dramatically raising the frequency of large impacts on the Earth. It is just a few million years now since our system last plunged through the galactic plane—could a phalanx of comets be heading for us at this very moment? By the time we find out it might very well be too late.

The Shiva hypothesis calls for a periodicity operating on truly geological timescales, and for this reason is rarely addressed in discussions of the immediate threat from impact events. Much more relevant to considerations of our own safety and survival—and that of our immediate descendants—is a proposal by UK astronomers Victor Clube and Bill Napier that the Earth is struck by clusters of objects every few *thousand* years, and that our planet took a serious pounding as recently as the Bronze Age—just 4,000 years ago. To find out what might cause such a worryingly recent bombardment we need to return to the Oort Cloud in deepest space. Leaving aside disturbance of the cloud due to the passage of the solar system around the galaxy, normality sees a new comet from the cloud every now and again falling in towards the inner solar system—maybe as frequently as every 20,000 years. The newcomer is rapidly 'captured' and torn apart by the strong gravitational fields of either the Sun or Jupiter, forming a ring of debris spread out along its orbit, but concentrated particularly around the position of the original comet itself. A large comet, broken up in this way, can 'seed' the inner solar system with perhaps a million 1-kilometre sized lumps of rock, dramatically increasing the numbers of Earth-threatening objects, and significantly raising the chances of our planet being hit. Clube, Napier, and others of this particular coherent catastrophist school propose that the last giant comet from the Oort Cloud entered our solar system towards the end of the last Ice Age—a mere 10,000 years or so ago—breaking up to form a mass of debris known as the *Taurid Complex*. Every December the Earth passes through part of this debris stream, resulting in the sometimes spectacular light show put on by the Taurid meteor storm, as small rocky fragments and gravel-sized stones burn up in the upper atmosphere. These innocuous bits and pieces only represent the tail end of the Taurid Complex, however, the heart of which contains a 5-kilometre wide Earth-crossing comet known as *Encke* and at least 40 accompanying asteroids any one of which would create global havoc if it struck our planet.

The distribution of debris along the Taurid Complex orbit about the Sun is rather like that of runners in a 10,000-metre race; while the majority are clustered together in a pack, the rest are dotted here and there around the track. Most years—according to the coherent catastrophists—the Earth's orbit crosses that of the Taurid Complex at a point where there is little debris, resulting in a pre-Christmas spectacle and little else. Every 2,500–3,000 years or so, the Earth passes through the equivalent of the runners' pack—and finds itself on the receiving end of a volley of rocky chunks perhaps up to 200–300 metres across. Benny Peiser, a social anthropologist at Liverpool's John Moores University, thinks that just such a bombardment around 4,000 years ago led to the fall of many early civilizations during the third millennium BC. He and others have interpreted contemporary accounts in terms of a succession of impacts, too small to have a global impact but quite sufficient to cause mayhem in the ancient worlds, largely through generating destructive atmospheric shock waves, earthquakes, tsunamis, and wildfires. Many urban centres in Europe, Africa, and Asia appear to have collapsed almost simultaneously around 2350 BC, and records abound of flood, fire, quake, and general chaos. These sometimes fanciful accounts are, of course, open to alternative interpretation, and hard evidence for bombardment from space around this time remains elusive. Having said this, seven impact craters in Australia, Estonia, and Argentina have been allocated ages of 4,000–5,000 years and the search goes on for others. Even more difficult to defend are propositions by some that the collapse of the Roman Empire and the onset of the Dark Ages may somehow have been triggered by increased numbers of impacts when the Earth last passed through the dense part of the Taurid Complex between 400 and 600 AD. Hard evidence for these is weak and periods of deteriorated climate attributed to impacts around this time can equally well be explained by large volcanic explosions. In recent years there has, in fact, been a worrying tendency amongst archaeologists, anthropologists, and historians to attempt to explain every historical event in terms of a natural catastrophe of some sort—whether asteroid impact, volcanic eruption, or earthquake—many on the basis of the flimsiest of evidence. As the aim of this volume is to shed light on how natural catastrophes can affect us all, I would be foolish to argue that past civilizations have not suffered many times at the hands of nature. Attributing everything from the English Civil War and the French Revolution to the fall of Rome and the westward march of Genghis Khan to natural disasters only serves, however, to devalue the potentially cataclysmic effects of natural hazards and to trivialize the role of nature in shaping the course of civilization.

How would you like to die?

If supporters of the Taurid Complex model are to be believed, and I should say now that their views remain very much in the minority amongst advocates of the impact threat, then we may have only another thousand years or so before a series of blinding flashes and crashing sonic booms heralds the arrival of the next batch of fragmented comet. Alternatively, we could face oblivion tomorrow or have to wait 50,000 years before a city is obliterated or the world plunged into cosmic winter beneath a cloud of pulverized rock. But whenever the skies next fall, how will it affect us? This will depend upon three things: (i) size of the object, (ii) how quickly it is travelling, and (iii) whether it hits the land or the ocean. Everything else being equal, the larger the impactor the more devastating and widespread will be its effects. To reiterate, a body in the 50–100-metre size range carries enough destructive power to wipe out a major city or a small European country or US state. The level and extent of associated devastation will increase progressively with larger impactors until the critical 1 kilometre size is reached. In addition to causing appalling destruction on a regional or sub-continental scale, the arrival of an object of this size will affect the entire planet through engendering a period of dramatic cooling and reduced plant growth. For impactors larger than 1 kilometre the effects on the planet's ecosystems become progressively more severe until mass extinctions wipe out a significant percentage of all species. The 10-kilometre object that struck the Earth off the Mexican coast at the end of the Cretaceous period, 65 million years ago, not only finished off the dinosaurs but also two-thirds of all species living at the time. Even more disturbingly, there is evidence of a major impact event at the end of the Permian period some 250 million years ago that left fewer than 10 per cent of species alive. In all, at least 7 out of 25 major extinctions in the geological record have been linked with evidence for large impacts, although as I mentioned in the previous chapter there is a school of thought that plays down the environmental effects of impact events and prefers to implicate huge outpourings of basalt lava in the great extinctions of the past.

The destructive potential of a chunk of rock hurtling into the Earth is directly related to the kinetic energy it carries, and this reflects not only the size of the object but also the velocity of the collision. Because they travel substantially faster, therefore, impacts by so-called *long-period* comets, whose orbits carry them far out into interstellar space, cause more destruction than either Earth-Crossing Asteroids or local comets that follow orbits confined to the heart of the solar system. Both the nature and scale of devastation also depends upon whether the impactor hits the land or the sea. Two-thirds of our planet's surface is covered by water, so statistically this is where the majority of asteroids and comets strike. In such cases, the amount of pulverized rock hurled into the atmosphere might be reduced, compared to a land collision. However, this small benefit is likely to be at least partly countered by the formation of giant tsunamis capable of wreaking havoc across an entire ocean basin.

Furthermore, the gigantic quantities of water and salt injected into the atmosphere may severely affect the climate and even temporarily damage the ozone layer. Most of the evidence for the environmental effects of impacts comes from studies of just two events, one small and the other enormous.

At the low end of the scale, in 1908 a small asteroid, estimated at around 50 metres across, penetrated the Earth's atmosphere and exploded less than 10 kilometres above the surface of Siberia in a region known as Tunguska. This huge blast, which expended roughly the energy equivalent of 800 Hiroshima atomic bombs, was heard over an area four times the size of the UK and flattened over 2,000 square kilometres of full-grown forest. The blast registered on seismographs thousands of kilometres distant and the atmospheric shock wave was picked up by barographs time and again as it travelled three times around the planet before dissipating. The gas and dust generated by the explosion led to exceptionally bright night skies over Europe, sufficient—according to one contemporary report—to allow cricket to be played in London after midnight. Because of its inaccessibility, the first Russian expedition did not reach Tunguska until a quarter of a century later, when Leonid Kulik and his team were perplexed by the absence of the huge crater they were expecting. Instead they found a circular patch of badly charred and flattened trees 60 kilometres across, caused by the airburst as the rock disintegrated explosively due to the huge stresses caused by entry into the atmosphere. As the region was sparsely inhabited, casualties due to the impact were small,

35 Trees flattened by the explosion in the atmosphere of a small asteroid over Tunguska (Siberia) in 1908

with perhaps a few killed and up to 20 injured, although reports are understandably sketchy. Four hours later, however, and the Earth would have rotated sufficiently to bring the great city of St Petersburg into the asteroid's range and the result would have been catastrophic.

The Tunguska events pale into insignificance when compared to what happened off the coast of Mexico's Yucatan Peninsula 65 million years earlier. Here a 10-kilometre asteroid or comet—its exact nature is uncertain—crashed into the sea and changed our world forever. Within microseconds, an unimaginable explosion released as much energy as billions of Hiroshima bombs detonated simultaneously, creating a titanic fireball hotter than the Sun that vaporized the ocean and excavated a crater 180 kilometres across in the crust beneath. Shock waves blasted upwards, tearing the atmosphere apart and expelling over a hundred trillion tonnes of molten rock into space, later to fall across the globe. Almost immediately an area bigger than Europe would have been flattened and scoured of virtually all life, while massive earthquakes rocked the planet. The atmosphere would have howled and screamed as *hypercanes* five times more powerful than the strongest hurricane ripped the landscape apart, joining forces with huge tsunamis to batter coastlines many thousands of kilometres distant.

Even worse was to follow. As the rock blasted into space began to rain down across the entire planet so the heat generated by its re-entry into the atmosphere irradiated the surface, roasting animals alive as effectively as an oven grill, and starting great conflagrations that laid waste the world's forests and grasslands and turned fully a quarter of all living material to ashes. Even once the atmosphere and oceans had settled down, the crust had stopped shuddering, and the bombardment of debris from space had ceased, more was to come. In the following weeks, smoke and dust in the atmosphere blotted out the Sun and brought temperatures plunging by as much as 15 degrees Celsius. In the growing gloom and bitter cold the surviving plant life wilted and died while those herbivorous dinosaurs that remained slowly starved. Life in the oceans fared little better as poisons from the global wildfires and acid rain from the huge quantities of sulphur injected into the atmosphere from rocks at the site of the impact poured into the oceans, wiping out three-quarters of all marine life. After years of freezing conditions the gloom following the so-called Chicxulub impact would eventually have lifted, only to reveal a terrible Sun blazing through the tatters of an ozone layer torn apart by the

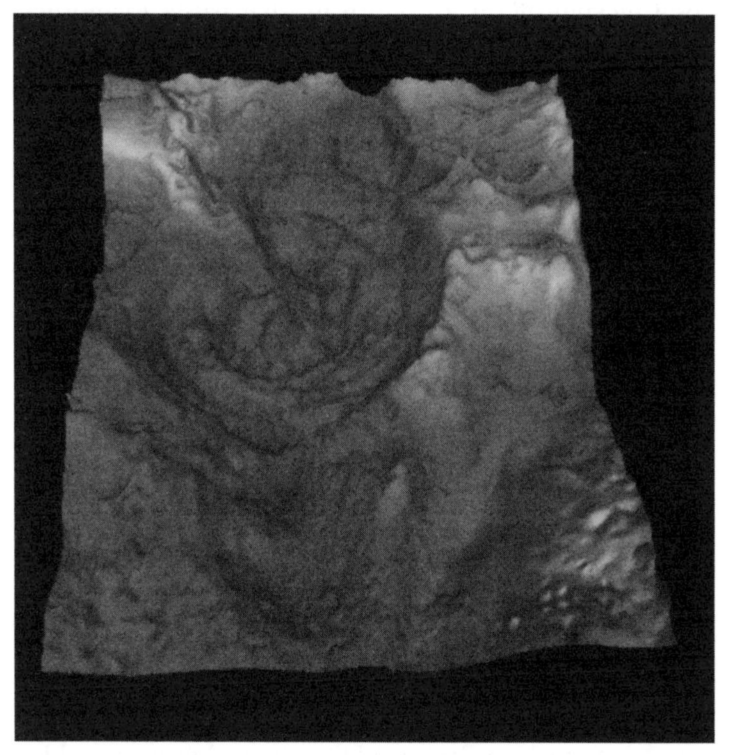

36 A reconstruction of the 65 million-year-old Chicxulub impact crater, now buried beneath younger rocks off Mexico's Yucatan Peninsula

chemical action of nitrous oxides concocted in the impact fireball: an *ultraviolet spring* hard on the heels of the cosmic winter that fried many of the remaining species struggling precariously to hang on to life. So enormously was the natural balance of the Earth upset that according to some it might have taken hundreds of thousands of years for the post-Chicxulub Earth to return to what passes for normal. When it did the age of the great reptiles was finally over, leaving the field to the primitive mammals—our distant ancestors—and opening an evolutionary trail that culminated in the rise and rise of the human race. But could we go the same way? To assess the chances, let me look a little more closely at the destructive power of an impact event.

At Tunguska, destruction of the forests resulted partly from the great heat generated by the explosion, but mainly from the blast wave that literally pushed the trees over and flattened them against the ground. The strength of this blast wave depends upon what is called the *peak overpressure,* that is the difference between ambient pressure and the pressure of the blast wave. In order to cause severe destruction this needs to exceed 4 pounds per square inch, an overpressure that results in wind speeds that are over twice the force of those found in a typical hurricane. Even though tiny compared with, say, the land area of London, the enormous overpressures generated by a 50-metre object exploding low overhead would cause damage comparable with the detonation of a very large nuclear device, obliterating almost everything within the city's orbital motorway. Increase the size of the impactor and things get very much worse. An asteroid just

250 metres across would be sufficiently massive to penetrate the atmosphere; blasting a crater 5 kilometres across and devastating an area of around 10,000 square kilometres—that is about the size of the English county of Kent. Raise the size of the asteroid again, to 650 metres, and the area of devastation increases to 100,000 square kilometres—about the size of the US state of South Carolina.

Terrible as this all sounds, however, even this would be insufficient to affect the entire planet. In order to do this, an impactor has to be at least 1 kilometre across, if it is one of the speedier comets, or 1.5 kilometres in diameter if it is one of the slower asteroids. A collision with one of these objects would generate a blast equivalent to 100,000 *million* tonnes of TNT, which would obliterate an area 500 kilometres across—say the size of England—and kill perhaps tens of millions of people, depending upon the location of the impact.

The real problems for the rest of the world would start soon after as dust in the atmosphere began to darken the skies and reduce the level of sunlight reaching the Earth's surface. By comparison with the huge Chicxulub impact it is certain that this would result in a dramatic lowering of global temperatures but there is no consensus on just how bad this would be. The chances are, however, that an impact of this size would result in appalling weather conditions and crop failures at least as severe as those of the 'Year Without a Summer', which followed the 1815 eruption of Indonesia's Tambora volcano. As mentioned in the last chapter, with even developed countries holding sufficient food to feed their populations for only a month or so, large-scale crop failures across the planet would undoubtedly have serious implications. Rationing, at the very least, is likely to be the result, with a worst case scenario seeing widespread disruption of the social and economic fabric of developed nations. In the developing world, where subsistence farming remains very much the norm, widespread failure of the harvests could be expected to translate rapidly into famine on a biblical scale. Some researchers forecast that as many as a quarter of the world's population could succumb to a deteriorating climate following an impact

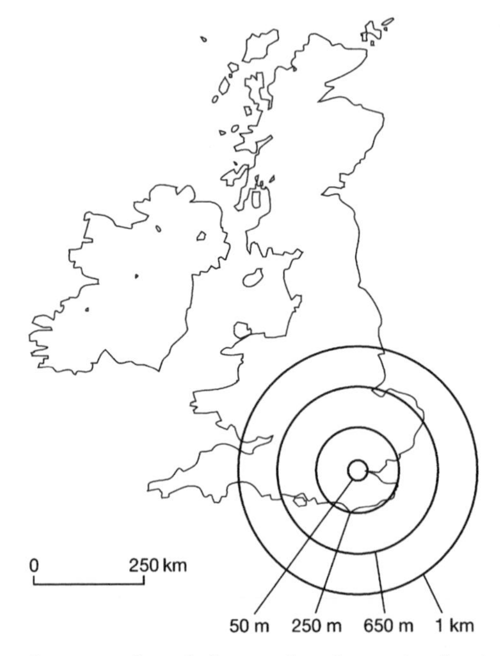

37 Predicted zones of total destruction for variously sized impacts centred on London

in the 1–1.5 kilometre size range. Anything bigger and photosynthesis stops completely. Once this happens the issue is not how many people will die but whether the human race will survive. One estimate proposes that the impact of an object just 4 kilometres across will inject sufficient quantities of dust and debris into the atmosphere to reduce light levels below those required for photosynthesis.

Because we still don't know how many threatening objects there are out there nor whether they come in bursts, it is almost impossible to say when the Earth will be struck by an asteroid or comet that will bring to an end the world as we know it. Impact events on the scale of the Chicxulub dinosaur-killer only occur every several tens of millions of years, so in any single year the chances of such an impact are tiny. Any optimism is, however, tempered by the fact that—should the Shiva hypothesis be true—the next swarm of Oort Cloud comets could even now be speeding towards the inner solar system. Failing this, we may have only another thousand years to wait until the return of the dense part of the Taurid Complex and another asteroidal assault. Even if it turns out that there is no coherence in the timing of impact events, there is statistically no reason why we cannot be hit next year by an undiscovered Earth-Crossing Asteroid or by a long-period comet that has never before visited the inner solar system. Small impactors on the Tunguska scale struck Brazil in 1931 and Greenland in 1997, and will continue to pound the Earth every few decades. Because their destructive footprint is tiny compared to the surface area of the Earth, however, it would be very bad luck if one of these hit an urban area, and most will fall in the sea. Although this might seem a good thing, a larger object striking the ocean would be very bad news indeed. A 500-metre rock landing in the Pacific Basin, for example, would generate gigantic tsunamis that would obliterate just about every coastal city in the hemisphere within 20 hours or so. The chances of this happening are actually quite high—about 1 per cent in the next 100 years—and the death toll could well top half a billion.

Estimates of the frequencies of impacts in the 1 kilometre size bracket range from 100,000 to 333,000 years, but the youngest impact crater produced by an object of this size is almost a million years old. Of course, there could have been several large impacts since, which either occurred in the sea or have not yet been located on land. Fair enough you might say, the threat is clearly out there, but is there anything on the horizon? Actually, there is. Some 13 asteroids—mostly quite small—could feasibly collide with the Earth before 2100. Realistically, however, this is not very likely as the probabilities involved are not much greater than 1 in 10,000—although bear in mind that these are pretty good odds. If this was the probability of winning the lottery then my local agent would be getting considerably more of my business. There is another enigmatic object out there, however. Of the 40 or so Near Earth Asteroids spotted last year, one—designated 2000SG344—looked at first as if it might actually hit us. The object is small, in the 100 metre size range, and its orbit is so similar to the earth that some have suggested it may be a booster rocket that sped one of the *Apollo* spacecraft on its way to the Moon. Whether hunk of rock or lump of man-made metal, it was originally estimated that 2000SG344 had a 1 in 500 chance of striking the Earth on 21 September 2030. Again, these may sound very long odds, but they are actually only five times greater than those recently offered during summer 2001 for England beating Germany 5–1 at football. We can all relax now anyway, as recent calculations have indicated that the object will not approach closer to the Earth than around five million kilometres. A few years ago, scientists came up with an index to measure the impact threat, known as the *Torino Scale*, and so far 2000SG344 is the first object to register a value greater than zero. The potential

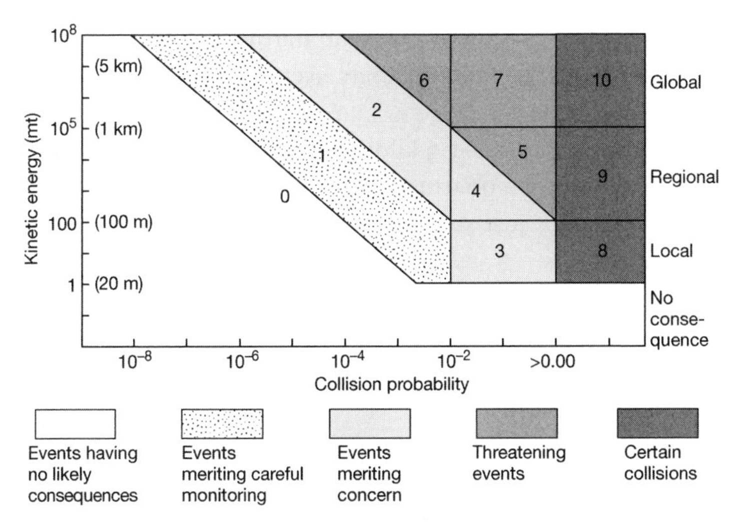

38 The Torino Scale classifies extra-terrestrial objects posing a threat to the planet in terms of their diameters, probability of impact, and degree of concern

impactor originally scraped into category 1, events meriting careful monitoring. Let's hope that many years elapse before we encounter the first category 10 event—defined as 'a certain collision with global consequences'. Given sufficient warning we might be able to nudge an asteroid out of the Earth's way but due to its size, high velocity, and sudden appearance, we could do little about a new comet heading in our direction.

Facts to fret over

- Perhaps 1,000 or more asteroids a kilometre or greater in diameter regularly cross the Earth's orbit and about a third of these will end their lives by colliding with our planet.
- Collision with a 1–1.5-kilometre impactor could result in the deaths of a quarter of the world's population.
- We may get just six months warning of a future comet impact.
- An asteroid large enough to destroy London, New York, or Paris strikes the planet a couple of times a century.
- Ancient impacts may have wiped out up to 90 per cent of all life on Earth.
- There is a 1 per cent chance of a 500-metre object hitting the Pacific Ocean in the next 100 years.
- According to some, we only have to wait another thousand years or so before our planet begins to take another pounding from space.

Epilogue

Now it's been ten thousand years
Man has cried a billion tears
For what he never knew
Now man's reign is through

In the year 2525 Zager & Evans

Pondering in isolation upon the consequences of global warming, the imminence of the next Ice Age, or the timing of a future super-eruption or asteroid impact can engender fleeting concern. Consideration *en masse* of the future threats to our planet and our race is quite capable, however, of contributing to bouts of severe depression. Let me summarize our current position. We are now well into a cycle of warming that is certain to lead to dramatic geophysical, social, and economic changes during the next hundred years that will impinge—probably largely detrimentally—on everyone. At the same time our planet is teetering on the edge of the next Ice Age, whose start global warming might actually bring forward, but which is likely to arrive within the next several thousand years even without our help. Asteroids sufficiently large to wipe out a quarter of the human race continue to hurtle across the Earth's orbit undetected, while who knows when we will detect the next great comet coming our way. There are now so many of us that the next giant tsunamis or volcanic super-eruption cannot fail to result in millions of deaths and the enormous disruption of our so-called advanced global society. So interconnected is our social and economic framework that just a single quake in Japan could lead to global economic disaster. There are other trends too that will ensure an end to the world as we know it, and within this century. Although the world's population is still rising, the rate started to slow in 1968 and numbers will peak at around 9 billion—half as great again as today—in about 2070. After that they will begin to fall and will be down to 8.4 billion just 30 years later. Welcome as this trend is, its legacy will be an ageing population and a so-called 'grey' future. By the end of the century an extraordinary 50 per cent or so of people in Japan and western Europe will be 60 or older, and a full third of the planet's population will be over this age. In a world where the elderly hold increasing sway over the young, some believe that intergenerational conflicts may come to dominate the political scene.

Not only will the future of our descendants be greyer, it will also be dull and barren. In one of the greatest mass extinctions ever, our activities are currently wiping out between 3,000 and 30,000 species a year, from an estimated total of just 10 million. Up to one-third of flowering plants could be at risk, while between 25 and 50 per cent of all animal species could disappear before the chimes ring in the new century on 1 January 2100. As more and more species are obliterated their places will quickly be taken by the pests, weeds, and diseases that live cheek-by-jowl with the human race. Instead of a world of gorillas, pandas, birds of paradise, and corals, our descendants will have to make do with rats, cockroaches, thistles, and nettles. Furthermore, biodiversity is such a fragile thing—tenderly and incrementally reared by evolution—that it may take 5 million years or more for it to restore itself. In the meantime, we will have committed an estimated 500 trillion of our descendants to life on a dull and—in terms of variety—largely lifeless planet. Just as importantly, we are actually playing God with evolution itself and the entire future prospects of life on Earth. By wiping out the bulk of species that exist today, we are destroying much of evolution's raw material and severely limiting the planet's ability to generate the species of the future. For millions of species throughout geological time the end of the world has already come, and our activities are ensuring that the same fate will shortly face many of the life forms with which we currently share our planet.

The picture I am painting of the future, then, just seems to get worse and worse. An ageing population, perhaps lacking the dynamism and innovation of the past, battling with climate change and its consequences and ensuing economic, political, and social upheaval, and struggling—maybe too late—to right the environmental wrongs wrought by their ancestors—us. Somehow, against this background Brandon Carter's 'doom soon' scenario does not sound so far-fetched. Others too have come to the conclusion that things must come to a dramatic head soon. Anders Johansen and Didier Sornette of the University of California, for example, have recently predicted—on the basis of trends in population and economic and financial indices—that some sort of abrupt switch to a new 'regime' will occur in 50 years or so, the nature of which remains to be seen, but which is unlikely to be pleasant.

Probably the only piece of good news that can be taken away from my brief look at the end of the world as we know it is that although this is going to happen—and soon—the survival of our race seems to be assured, for now at least. Leaving aside the possibility of a major comet or asteroid impact on a scale of the dinosaur-killer 65 million years ago—which only happen every few hundred million years—it is highly unlikely that anything else is going to wipe out every single last one of us—all 6 billion plus—in the foreseeable future. Even the replacement of the world with which we have become so familiar with one of sweltering heat or bitter cold might not seem as scary for those of our descendants likely to be in the thick of things. After all, we are a remarkably adaptable species, and can change to match new circumstances with some aplomb. Familiar 'worlds' have certainly ended many times before, as no doubt a centenarian born and raised while Queen Victoria sat on the throne of the United Kingdom, and who lived to see man land on the moon, would testify. The danger is, however, that the world of our children and those that follow will be a world of struggle and strife with little prospect of, and perhaps little enthusiasm for, progress as the Victorians viewed it. Indeed, it would not be entirely surprising if, at some future time, as the great coastal cities sink beneath the waves or below sheets of ice, the general consensus did not hold that there had been quite enough progress thank you—at least for a while.

While I have tried in these pages to extrapolate current trends and ideas to tease out and examine somewhat depressing scenarios for the future of our planet and our race, I am sure that, to some extent at least, you would be justified in accusing me of a failure of the imagination. After all, I have rarely looked ahead beyond a few tens of thousands of years, and yet our Sun will still be bathing our planet in its life-giving warmth for another 5 billion years or more. Who knows, over that incomprehensible length of time, what *Homo sapiens* and the species that evolve from us will do and become. Our species and those that follow may be knocked back time and time again in the short term, but provided we learn to nurture our environment rather than exploit it, both here on Earth—before the Sun eventually swallows it up—and later, perhaps, in the solar system and the galaxy and beyond, then we have the time to do and be almost anything. Maybe now is the right time to start.

Appendix A

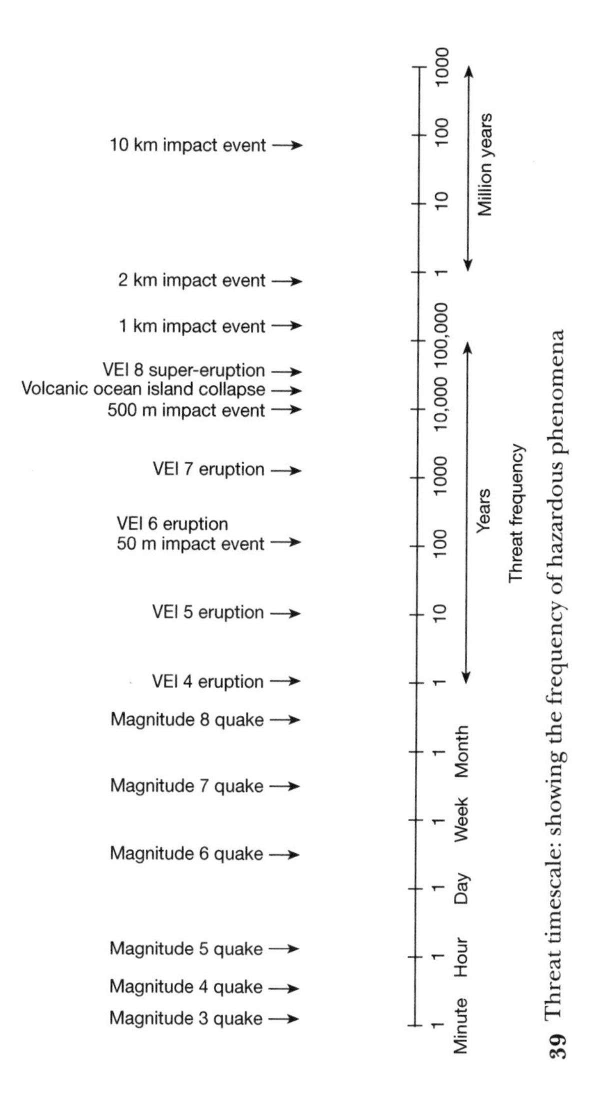

39 Threat timescale: showing the frequency of hazardous phenomena

Appendix B

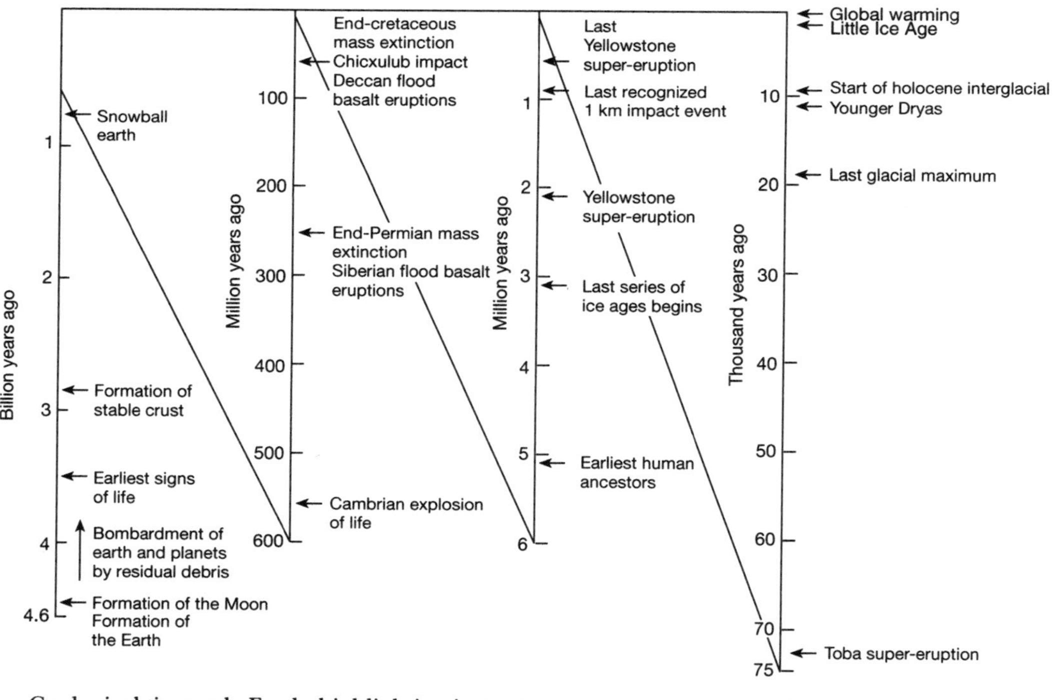

40 Geological timescale Earth: highlighting important events mentioned in this book

Further Reading

Alexander, David. Natural Disasters. UCL Press. 1998.

Bryant, Edward. Tsunami: The Underrated Hazard. Cambridge University Press. 2001.

Burroughs, William. J. Climate Change. Cambridge University Press. 2001.

Dawson, Alastair. G. Ice Age Earth. Routledge. 1992.

Intergovernmental Panel on Climate Change (IPCC) Working Group II. Climate Change 2001: Impacts, Adaptation, and Vulnerability. Cambridge University Press. 2001.

Lamb, H. H. Climate, History, and the Modern World. Routledge. 1997.

Legett, Jeremy. The Carbon War. Penguin. 2000.

Lovelock, James. Gaia: The Practical Science of Planetary Medicine. Gaia. 1991.

McGuire, Bill. Apocalypse: A Natural History of Global Disasters. Cassell. 1999.

McGuire, Bill. Raging Planet: Earthquakes, Volcanoes and the Tectonic Threat to Life on Earth. Barrons. 2002.

McGuire, W. J., Kilburn, C. R. J., and Mason, I. M. Natural Hazards and Environmental Change. Arnold. 2001.

Newson, L. The Atlas of the World's Worst Natural Disasters. Dorling Kindersley. 1998.

Steel, Duncan. Target Earth. Readers Digest. 2000.

Zebrowski, E. Jr. Perils of a Restless Planet: Scientific Perspectives on Natural Disasters. Cambridge University Press. 1997.

Index